Harold M. Edwards

Linear Algebra

1995
Birkhäuser

Harold M. Edwards
Courant Institute
New York University
New York, NY 10012

Library of Congress Cataloging In-Publication Data

Edwards, Harold M.
Linear algebra / Harold M. Edwards.
p. cm.
ISBN 0-8176-3731-1. -- ISBN 3-7643-3731-1 (Basel)
1. Algebra, Linear. I. Title.
QA184.E355 1994 94-35356
512.9'43--dc20 CIP

Printed on acid-free paper

Birkhäuser 

ISBN 0-8176-3731-1
ISBN 3-7643-3731-1

Layout, design, and macros by Martin Stock, Cambridge, MA
Typeset by TEXniques, Inc., Brighton, MA
Printed and bound by Quinn-Woodbine, Woodbine, NJ
Printed in the U.S.A.

9 8 7 6 5 4 3 2 1

Contents

8 Matrices with Polynomial Entries

9 Similarity of Matrices

10 The Spectral Theorem

Appendix Linear Programming

Preface

Although the title "Linear Algebra" describes the place of this book in the mathematics curriculum, a better description of its actual content would be "The Arithmetic of Matrices." The ability to compute with matrices—even better, the ability to *imagine* computations with matrices—is useful in any pursuit that involves mathematics, from the purest of number theory to the most applied of economics or engineering. The goal of this book is to help students acquire that ability.

Accordingly, the emphasis throughout the book is on algorithms. A by-product of this emphasis is the complete disappearance of set theory, a disappearance that will greatly disturb teachers accustomed to the standard linear algebra course but will not, and should not, disturb students in the least. The material in the book will be helpful, in addition to its other uses, in learning the language and the peculiar habits of mind that are set theory (see the Supplementary Unit of Chapter 6). The standard linear algebra course attempts to reverse the order and to use set theory to teach linear algebra—an approach that is as silly as it is unsuccessful.

Each chapter has an "Examples" section giving sample applications of the algorithms contained in the chapter. The student should refer to the examples while reading the chapter. The chapter necessarily deals with the most general case, but no mathematician ever understands the most general case first. One famous mathematician, David Hilbert, gave his students the excellent advice, "Always begin with the simplest example."

I mentioned *imaginary* calculation in the first paragraph. The emphasis on algorithms in this course lends itself very well to the use of computers, and I hope that teachers will be able to take advantage of local computer facilities and the particular levels of computer competence their students have to go further with the computational aspect of the subject. However, I do believe that the main thing is to be able to *imagine* the computation. The best theorems of mathematics have the form, "If you do calculation X, you will find result Y." To understand the meaning of such a theorem you must be able to imagine the calculation, but the actual calculation is rarely needed; if it *is* needed, you need to be able to imagine alternative ways of doing it in order to select a good one. The main reason for doing computations in a linear algebra course is to develop this form of imagination. It is not altogether clear that doing large examples on a computer is better for this purpose than doing small examples with pencil and paper. It is best, no doubt, to do both.

Some of the proofs, particularly in the last few chapters, involve lengthy arguments. My answer to the perennial question 'Are we expected to know the proofs for the exam?' is 'no,' not because proofs are unimportant, but because a proof cannot be *known,* it must be *understood,* and understanding is very hard to test on an exam. The most natural exam question is a particular example that uses the theorem. Preparing for such a question is the best way to study the material of the course, because there is no way to work through an example without referring to the proof of the most general case, just as there is no way to grasp the general case covered by the proof without applying it to specific examples. After enough work with specific examples, the meaning of the proof should begin to become clear.

Synopsis

The first algorithm of matrix arithmetic is matrix multiplication. As Chapter 1 explains, this operation arises naturally as the operation of composition of linear substitutions. The naturalness of the idea is confirmed by its appearance in a wide variety of apparently disparate contexts in mathematics.

Matrix multiplication is such a simple operation that it scarcely deserves to be called an algorithm. The same can certainly not be said of matrix division. If a matrix product $AX = Y$ is known, and if one of the factors A is known, to what extent is the other factor X determined, and how can all possible Xs be found? The algorithmic answer to this question in Chapter 3 depends on the notion of matrix equivalence developed in Chapter 2.

In the definition of Chapter 2, a matrix is equivalent to another if it can be transformed into the other by a sequence of operations of four elementary types: Adding a column to an adjacent column, adding a row to an adjacent row, subtracting a column from an adjacent column, or subtracting a row from an adjacent row. As is proved by an algorithm in Chapter 2, every matrix is equivalent to a diagonal one. The solution of the division problem consists of two simple observations; first, that the division problem $AX = Y$ is easy to solve when A is diagonal; second, that solution of the division problem for one matrix A leads easily to its solution for any equivalent matrix.

The algorithm of Chapter 2 gives a method of finding equivalences between matrices, but it does not give a method for proving that matrices are *inequivalent.*

Chapter 5 gives an algorithm for determining whether two given matrices are equivalent by giving a *canonical form* for matrices under equivalence—that is, a form such that any matrix is equivalent to one in canonical form, and such that two matrices in canonical form are equivalent only if they are equal.

The proof in Chapter 5 that matrices in canonical form are equivalent only if they are equal requires the theory of determinants. For this reason, Chapter 4 deals with determinants. A determinant of a square matrix is defined here to be the product of the diagonal entries of an equivalent diagonal matrix. The main theorem of Chapter 4 states that *a square matrix has only one determinant.* The proof is algorithmic in the sense that it depends on a formula—a very concrete algorithm, although in this case usually an impractical one—expressing the determinant in a way that does not depend on finding an equivalent diagonal matrix.

The first five chapters deal with matrices with *integer* entries. Not only do these matrices exhibit all the interesting properties that deserve to be covered, they also exhibit some properly number-theoretic or "arithmetic" properties that matrices with rational entries lack, namely, the *divisibility* conditions for the solvability of $AX = Y$, conditions which disappear when rational solutions are allowed. For example, in the first five chapters the equation $2x = y$ is considered to have a solution when y is even but no solution when y is odd. In addition to preserving this important feature of the arithmetic of matrices, the restriction to integer entries makes the computations easier to imagine.

Chapter 6 deals with the arithmetic of matrices with rational entries. Briefly put, it shows that *nothing changes* in the transition from integer to rational entries, except that there may be additional solutions to division problems. In particular, the inversion problem $AX = I$ has a solution—that is, A has a right inverse—whenever A is not a right zero divisor. Similarly, A has a left inverse whenever it is not a left zero divisor.

The method of least squares and its generalization, the theory of Moore-Penrose generalized inverses, are the subjects of Chapter 7. The literature does not seem to contain a less cumbersome name for Moore-Penrose generalized inverses, so I have taken the liberty of giving them the monosyllabic and rather descriptive name "mate." Every $m \times n$ matrix of rational numbers has one and only one mate, which is an $n \times m$ matrix of rational numbers. The mate B of a matrix A provides an approximate solution X of $AX = Y$ by means of the formula $X = BY$. What qualifies $X = BY$ to be called the best solution of $AX = Y$ is that (1) the sum of the squares of the entries of the matrix $AX - Y$, which is denoted $\|AX - Y\|^2$, achieves its minimum when $X = BY$, and (2) among all matrices X for which $\|AX - Y\|^2$ achieves its minimum, BY is the one for which $\|X\|^2$ is smallest.

Two square matrices A and B are called *similar* if there is an invertible square matrix P such that $B = P^{-1}AP$. *Two matrices A and B are similar if and only if $xI - A$ and $xI - B$ are equivalent as matrices of polynomials in one variable.* Here, in a natural analogy to the notion of equivalence of matrices in the first five chapters, two matrices of polynomials in x are said to be equivalent if one can be transformed into the other by a sequence of operations in which a multiple of a row is added to an adjacent row or a multiple of a column is added to an adjacent column. This definition is introduced in Chapter 8 and, in a manner that is completely analogous to Chapter 5, an algorithm is given for determining whether two matrices of polynomials are equivalent.

Chapter 9 then deals with similarity of matrices and, by means of the theorem stated above and the algorithm for deciding whether two matrices of polynomials are equivalent, gives an algorithm for deciding whether two square matrices are similar. This point of view leads naturally to the minimum polynomial, the elementary divisors, and the rational canonical form of a square matrix. Jordan canonical form is more familiar today than rational canonical form. The Jordan form requires the use of *complex numbers,* but the rational form uses only rational numbers; when complex numbers are allowed, the rational form becomes the Jordan form.

The final chapter, Chapter 10, is devoted to the spectral theorem for symmetric matrices, otherwise known as the principal axes theorem. An *orthogonal partition of unity* is a set of nonzero symmetric matrices $P_1, P_2, \ldots, P_k$ for which $P_1 + P_2 + \cdots + P_k = I$, $P_i^2 = P_i$, and $P_iP_j = 0$ when $i \neq j$. Given an orthogonal partition of unity $P_1 + P_2 + \cdots + P_k = I$ and given a list $a_1, a_2, \ldots, a_k$ of distinct numbers, $a_1P_1 + a_2P_2 + \cdots + a_kP_k = S$ is a symmetric matrix. Loosely speaking, the spectral theorem states that every symmetric matrix S is of this form for one and only one orthogonal partition of unity $P_1 + P_2 + \cdots + P_k = I$ and set of numbers $a_1, a_2, \ldots, a_k$.

For this statement to be true, however, the notion of "number" must be enlarged from the algebraic notion of "rational number" used throughout the book to the notion of "real number." Real numbers involve *limits* and therefore lie outside the domain of algebra. Chapter 10 stays within this domain and proves two purely algebraic statements, from which the full theorem follows easily once the theory of real numbers is available. These theorems are that (1) if the roots of the characteristic polynomial of a symmetric matrix S are all rational, then S can be written in just one way in the form $S = a_1P_1 + a_2P_2 + \cdots + a_kP_k$, and (2) for any symmetric matrix S with rational (or real) entries, the number of sign changes of the minimum polynomial is equal to its degree. The second of these theorems implies that the roots of the characteristic polynomial of a symmetric matrix are *real* numbers, from which it follows that the algorithm of theorem (1) for constructing $P_1 + P_2 + \cdots + P_k = I$ and $a_1, a_2, \ldots, a_k$ with $S = a_1 P_1 + a_2 P_2 + \cdots + a_k P_k$ can be used to express S in the desired form, except that the as and the entries of the matrices P are real, not necessarily rational, numbers.

The Appendix gives a terse but full description of the two-phase simplex method, which is an algorithm for finding the solution of an arbitrary linear programming problem.

Chapter 1
Matrix Multiplication

1 Linear Substitutions

THE POINT OF VIEW adopted in this book is that the main subject of linear algebra is **linear substitutions**, that is, recipes for expressing one set of variables as sums of multiples of another set of variables. In the first five chapters, the multipliers of the linear substitutions will be taken to be *integers.* The substitutions are called "linear" because an expression which involves a sum of multiples of variables is called a *homogeneous linear* polynomial,* which is to say, a polynomial in which each term contains exactly one variable and that variable occurs to the first power. For example,

$$\begin{aligned} u &= 2x + 3y - z \\ v &= x - y + 5z \end{aligned} \tag{1}$$

and

$$\begin{aligned} x &= -a + 4b \\ y &= 2a + 3b \\ z &= 7a - 5b \end{aligned} \tag{2}$$

are linear substitutions. More generally, a linear substitution involves two sets of variables, which we will call the *original* variables and the *new* variables, and equations telling how to replace each of the original variables with an expression involving the new variables.

The "variables" in a linear substitution are more properly called "indeterminates"; they are mere placeholders in an algebraic manipulation and it is not necessary to regard them as representing variable numbers. However, because the word "indeterminate" is so cumbersome, the familiar and only slightly misleading term "variable" will be used.

Readers may be dismayed to learn that the main subject of the book is the "algebraic manipulation" of "placeholders," but they should not be. After all,

*It is often stated that such a polynomial is called "linear" because the equation of a line involves a linear polynomial. However, the historical reason appears to be the same reason that a polynomial involving terms of degree 2 is called "quadratic" and a polynomial of degree 3 is called "cubic." The word "linear" in the title of this book therefore means "of degree one" or "no powers higher than the first."

ordinary arithmetic is nothing more than the manipulation of integers—a sort of high-powered counting—and no one can doubt that arithmetic is worth learning. Linear algebra can be regarded as the arithmetic of linear substitutions, and its usefulness in dealing with questions in economics, engineering, physics, pure and applied mathematics, and many other fields is quite comparable to that of arithmetic.

2 *Composition of Linear Substitutions*

As the name implies, the primary function of a linear substitution is that it can be *substituted* in algebraic expressions involving the original variables to produce new algebraic expressions involving the new variables.

If the algebraic expressions in which the substitution is performed describe a linear substitution, then the new algebraic expressions also describe a linear substitution. For example, if the linear substitution (2) of Section 1 is substituted in the substitution (1), the result is

$$\begin{aligned} u &= 2(-a+4b) + 3(2a+3b) - (7a-5b) \\ &= (-2+6-7)a + (8+9+5)b = -3a+22b \\ v &= (-a+4b) - (2a+3b) + 5(7a-5b) \\ &= (-1-2+35)a + (4-3-25)b = 32a-24b, \end{aligned}$$

that is,

$$\begin{aligned} u &= -3a + 22b \\ v &= 32a - 24b. \end{aligned} \tag{3}$$

The operation of substituting one linear substitution in another is called **composition** of linear substitutions. It is meaningful, of course, only for ordered pairs of linear substitutions in which the new variables of the first coincide with the original variables of the second; in the composed linear substitution, as in (3), the "original" variables are the original variables of the first substitution (1) and the "new" variables are the new variables of the second substitution (2).

Linear algebra is to a very large extent the study of properties of the operation of composition of linear substitutions. In particular, a key technique of this book is the technique in Chapter 2 of writing a given linear substitution as a composition of many simple linear substitutions—namely, as a composition of a simple type of linear substitutions that we will call "tilts" and of one linear substitution that is "diagonal."

3 Matrices

The choice of good notation is one of the most important aspects of the development of any mathematical subject. In the dispute over whether Newton or Leibniz was the discoverer of the calculus, few doubt that Newton was the more powerful mathematician; on the other hand, Leibniz's long-term influence on the development of the subject was greater because of his scrupulous attention to the matter of notation and because of the wonderful notation of differentials and integrals that he devised.

It is important when choosing notation to consider carefully what to leave out. Clear thinking requires a lack of ambiguity about concepts, which suggests stating things very explicitly and at length; on the other hand, it is easy to become muddled in both thinking and computation by too much detail and too much repetition. An effective compromise, in which there is enough explicitness to avoid ambiguity but not so much that it gets in the way, is necessary.

In a linear substitution, the information about the substitution is contained mainly in its *matrix of coefficients,* the rectangular array of numbers which are its multipliers. In the cases of the linear substitutions (1), (2), and (3) above, these matrices* are

$$\begin{bmatrix} 2 & 3 & -1 \\ 1 & -1 & 5 \end{bmatrix}, \quad \begin{bmatrix} -1 & 4 \\ 2 & 3 \\ 7 & -5 \end{bmatrix}, \text{ and } \begin{bmatrix} -3 & 22 \\ 32 & -24 \end{bmatrix}, \tag{4}$$

respectively. In fact, the matrix of coefficients completely describes the corresponding linear substitution, except that it does not give the names of its variables. For example, the 2×3 matrix† which is the first one in the list (4) above describes a linear substitution in which two original variables are replaced by sums of multiples of three new variables; when the two original variables are named u and v and the three new variables are named x, y, and z, the linear substitution in question is (1) of Section 1.

In many situations the names of the variables can be inferred from the context so that the matrix of coefficients is all that needs to be written. In many other situations the names of the variables are of no interest, and in fact, in the case of intermediate substitutions, it may not even be necessary to *give* names to the variables of a substitution. Therefore, a linear substitution is, for practical purposes, described completely by its matrix of coefficients.

*Note that the plural of "matrix" is "matrices," pronounced MATRI-SEEZE. It is easier for most students to use the alternative plural "matrixes," which is not incorrect, but mathematicians normally use the more classical plural form.

†In describing the dimensions of a matrix, one always gives first the number of *rows* or horizontal lines of numbers, in this case 2, and then the number of *columns* or vertical lines of numbers, in this case 3.

It is stated at the end of Section 1 that linear algebra can be regarded as the arithmetic of linear substitutions. Because matrices and linear substitutions are effectively the same, and because it is with matrices that one normally computes, the presentation given in this book is in fact a presentation of linear algebra as the arithmetic of *matrices,* and the central concept of a linear substitution will hardly be mentioned in later chapters. However, a matrix is simply a rectangular array of numbers; only when the computations of linear algebra are regarded as involving matrices of coefficients of linear substitutions do these computations have a clear meaning. In particular, only in terms of linear substitutions does the operation of matrix multiplication have a clear meaning—it is the *composition* of two linear substitutions.

4 *Matrix Multiplication*

Matrix multiplication is *by definition* the operation on matrices which describes the composition of linear substitutions. Specifically, let M_1 be an $m \times n$ matrix and let M_2 be an $n \times k$ matrix. Then M_1 can be regarded as the matrix of coefficients of a linear substitution in which there are m original variables and n new variables; similarly, M_2 can be regarded as the matrix of coefficients of a linear substitution with n original variables and k new variables. Let the variables in these two linear substitutions be named in such a way that the n new variables in the first substitution have the same names, in the same order, as the n original variables in the second. Then the composition of the two linear substitutions is a well-defined linear substitution with m original variables and k new variables, so it is represented by $m \times k$ matrix. This $m \times k$ matrix is by definition the product of M_1 and M_2, and is denoted M_1M_2.

Note that a matrix product M_1M_2 is defined if and only if the number of columns of M_1 is equal to the number of rows of M_2. (The new variables of the substitution represented by M_1 must be the same as the original variables of the substitution represented by M_2.) If the number of columns of M_1 is *not* equal to the number of rows of M_2, then it is meaningless to talk of the product of M_1 and M_2. (This implies, by the way, that M_2M_1 may very well be meaningless even when M_1M_2 is defined. Therefore, there is certainly no analog in matrix multiplication for the commutative law of multiplication of numbers.)

Note also that the number of rows of M_1M_2 is equal to the number of rows of M_1 (because the original variables of the first substitution are the original variables of the composed substitution) and that the number of columns of M_1M_2 is equal to the number of columns of M_2 (because the new variables of the second substitution are the new variables of the composed substitution).

5 The Computation of Matrix Products

The rule for the computation of matrix products is easy to derive. Consider, for example, the composition of the linear substitutions

$$\begin{aligned} r &= 2m - n \\ s &= m - 2n \\ t &= 3m + 3n \\ u &= -m + 5n \end{aligned} \quad \text{and} \quad \begin{aligned} m &= 3f + 2g - 2h \\ n &= -f - 3g + 5h \end{aligned} \tag{5}$$

which a brief computation shows to be

$$\begin{aligned} r &= 7f + 7g - 9h \\ s &= 5f + 8g - 12h \\ t &= 6f - 3g + 9h \\ u &= -8f - 17g + 27h. \end{aligned} \tag{6}$$

As a matrix multiplication, this composition is represented by the equation

$$\begin{bmatrix} 2 & -1 \\ 1 & -2 \\ 3 & 3 \\ -1 & 5 \end{bmatrix} \begin{bmatrix} 3 & 2 & -2 \\ -1 & -3 & 5 \end{bmatrix} = \begin{bmatrix} 7 & 7 & -9 \\ 5 & 8 & -12 \\ 6 & -3 & 9 \\ -8 & -17 & 27 \end{bmatrix}. \tag{7}$$

The rule by which the matrix on the right is determined by the two matrices on the left can be found as follows. Consider the entry -17 in the middle of the last row of the matrix on the right. This -17 is the coefficient of g in the expression for u in the composed linear substitution (6). It is clear that this -17 is independent of the entries in the first three rows of M_1 because these rows relate only to the variables r, s, and t, not to u. Similarly, the -17 is independent of the entries in the first and last columns of M_2, because these columns of M_2 relate only to the variables f and h. So the question of how the entry -17 is determined by the entries in M_1 and M_2 reduces to the question of how it is determined by the two entries in the last row of M_1 and the two entries in the middle column of M_2. An examination of the computation of the composed linear substitution shows that it comes from the computation $-17 = (-1) \cdot 2 + 5 \cdot (-3)$, where the first term $(-1) \cdot 2$ is the product of the first entries in the row and column in question—that is, the product of the entries relating to the variable m—and the second term $5 \cdot (-3)$ is the product of the second entries, those relating to the variable n. Once these observations have been made, it is easy to see that the other entries in M_1M_2 in formula (4) are found in the same way. For example, the second entry in the third column of M_1M_2 is $-12 = 1 \cdot (-2) + (-2) \cdot 5$, the number found using the second row of M_1 and the last column of M_2 and adding the product of the first entries to the product of the second entries.

The general case is clearly the same. If M_1 has m rows and n columns and if M_2 has n rows and k columns, the entry in the ith row and the jth column of $M_1 M_2$ is found by using just the ith row of M_1 and the jth column of M_2 and adding the n products of corresponding entries. Note that the composition of the linear substitutions (1) and (2) of Section 1 to find (3) of Section 2, a composition which in matrix form is expressed by the equation

$$\begin{bmatrix} 2 & 3 & -1 \\ 1 & -1 & 5 \end{bmatrix} \begin{bmatrix} -1 & 4 \\ 2 & 3 \\ 7 & -5 \end{bmatrix} = \begin{bmatrix} -3 & 22 \\ 32 & -24 \end{bmatrix},$$

can easily be found using this method. For example, the entry in the first row of the first column is $2 \cdot (-1) + 3 \cdot 2 + (-1) \cdot 7 = -3$.

The operation of matrix multiplication (that is, the operation of composition of linear substitutions) is absolutely of fundamental importance in linear algebra. The reader should practice with the examples at the end of the chapter to develop the ability to do it easily. Run the index finger of your left hand across the row of the first factor and the index finger of your right hand down the column of the second factor, calculating the products and adding them in your head as you go.

6 *Associativity*

As was noted in Section 4, the commutative law of multiplication makes no sense for the operation of matrix multiplication. However, the *associative* law of multiplication *does* apply to matrix multiplication. Specifically, if three matrices M_1, M_2, and M_3 are given, and if $M_1 M_2$ and $M_2 M_3$ are both defined, then $(M_1 M_2) M_3$ and $M_1 (M_2 M_3)$ are both defined, and they are equal.

The associativity of matrix multiplication is a consequence of the more general law of associativity of algebraic substitutions (not just linear substitutions). If a string of three substitutions are such that the composition of the first with the second and the composition of the second with the third are both defined, then the composition of *all three* is defined and there is no need to specify which of the substitutions is performed first. For example, if $w = z^2$, $z = y + 1$, and $y = 2x$, then w can be expressed in terms of x either by first finding $w = (y + 1)^2$ and then $w = (2x + 1)^2$ or by first finding $z = 2x + 1$ and then $w = (2x + 1)^2$.

The truth of the associative law for substitutions is such a simple computational fact that it almost goes without saying. However, in the special case of the associative law of matrix multiplication, it is one of the most important facts of linear algebra.

Examples

The matrix product $\begin{bmatrix} 1 & 2 \\ 3 & 4 \end{bmatrix}\begin{bmatrix} -1 & 1 \\ 2 & -2 \end{bmatrix} = \begin{bmatrix} 3 & -3 \\ 5 & -5 \end{bmatrix}$ is an abbreviation of the composition of two linear substitutions, say

$$\begin{aligned} a &= c + 2d \\ b &= 3c + 4d \end{aligned} \qquad \text{and} \qquad \begin{aligned} c &= -e + f \\ d &= 2e - 2f \end{aligned}$$

to find

$$\begin{aligned} a &= (-e + f) + 2(2e - 2f) = 3e - 3f \\ b &= 3(-e + f) + 4(2e - 2f) = 5e - 5f. \end{aligned}$$

Note that when the variables are named there is no tendency to confuse this matrix product with the product of the same matrices in the reverse order

$$\begin{bmatrix} -1 & 1 \\ 2 & -2 \end{bmatrix}\begin{bmatrix} 1 & 2 \\ 3 & 4 \end{bmatrix} = \begin{bmatrix} 2 & 2 \\ -4 & -4 \end{bmatrix}$$

which represents an entirely different composition of linear substitutions. A third linear substitution $e = -g$ and $f = 2g$ can be composed with the two composed above to give

$$\begin{aligned} a &= 3(-g) - 3(2g) = -9g \\ b &= 5(-g) - 5(2g) = -15g. \end{aligned}$$

The expression of a and b in terms of g can also be found by first expressing c and d in terms of g, namely,

$$\begin{aligned} c &= -(-g) + (2g) = 3g \\ d &= 2(-g) - 2(2g) = -6g \end{aligned}$$

and then substituting in the expression of a and b in terms of c and d to find

$$\begin{aligned} a &= (3g) + 2(-6g) = -9g \\ b &= 3(3g) + 4(-6g) = -15g. \end{aligned}$$

In terms of matrix multiplication, the fact that these two ways of finding the expression of a and b in terms of g give the same result is the statement of the associative law of multiplication

$$\left(\begin{bmatrix} 1 & 2 \\ 3 & 4 \end{bmatrix}\begin{bmatrix} -1 & 1 \\ 2 & -2 \end{bmatrix}\right)\begin{bmatrix} -1 \\ 2 \end{bmatrix} = \begin{bmatrix} 1 & 2 \\ 3 & 4 \end{bmatrix}\left(\begin{bmatrix} -1 & 1 \\ 2 & -2 \end{bmatrix}\begin{bmatrix} -1 \\ 2 \end{bmatrix}\right).$$

Because of the associative law of composition of substitutions—or, what is the same, the associative law of matrix multiplication—a product of three or more matrices can be written *without parentheses.*

Exercises

1 Find the composition of

$$\begin{aligned} u &= 3x + 2y \\ v &= 7x - 5y \\ w &= -x + 11y \end{aligned} \qquad \text{and} \qquad \begin{aligned} x &= a + b + c \\ y &= 3a - 2b + 4c \end{aligned}$$

writing your answer both as a direct substitution and as a matrix multiplication.

2 Do the same for

$$\begin{aligned} p &= u + v + w \\ q &= 3u - 2v + 4w \end{aligned} \qquad \text{and} \qquad \begin{aligned} u &= 3x + 2y \\ v &= 7x - 5y \\ w &= -x + 11y. \end{aligned}$$

3 Find the composition of the three linear substitutions in the first two exercises.

4 Find the composition of the linear substitutions

$$\begin{aligned} u &= 3x + z \\ v &= x + 2y - z \end{aligned}$$

and

$$\begin{aligned} x &= 3a + b \\ y &= 2b - c \\ z &= a + b + c. \end{aligned}$$

5 A *square* matrix—one with the same number of rows as columns—can be multiplied by itself, and this operation can be applied repeatedly to find *powers* of the matrix. Show that if $J = \begin{bmatrix} 0 & -1 \\ 1 & 0 \end{bmatrix}$ and if n is an integer greater than 4, then $J^n = J^{n-4}$.

6 (a) With $A = \begin{bmatrix} 1 & 1 \\ 0 & 1 \end{bmatrix}$ find, by computing a few examples, a formula for A^n where $n = 1, 2, 3, \ldots$. (b) Write a linear substitution whose matrix of coefficients is A, and *invert* this substitution. (c) Relate the answer for (a) to the answer for (b).

7 Let $A = \begin{bmatrix} 1 & 0 & 0 \\ 0 & 1 & 1 \\ 0 & 0 & 1 \end{bmatrix}$. The product AB is defined for any matrix B with 3 rows. Compute AB for a few sample matrices with 3 rows and formulate a rule for finding AB when B is given. Similarly, BA is defined whenever B has 3 columns. How is BA found when B is given?

8 (a) Find a 4×4 matrix A which has the property that multiplication of a matrix with 4 rows on the left by A adds the second row to the third while leaving the first, second, and fourth rows unchanged; that is, for any B with four rows, B and AB are the same in rows 1, 2, and 4, but row 3 of AB is the sum of rows 2 and 3 of B. (b) Find a 4×4 matrix A which has the property that multiplication of a matrix with 4 columns on the right by A adds the second column to the third while leaving the first, second, and fourth columns unchanged.

9 Find the inverses of the two matrices of the preceding exercise, that is, the linear substitutions which undo what these linear substitutions do. (See Section 1 of Chapter 2.)

10 Find all products of two of the matrices

$$A = \begin{bmatrix} 1 \\ 2 \end{bmatrix}, \quad B = [3], \quad C = [4 \ \ 5 \ \ 6], \quad D = \begin{bmatrix} 0 & 1 \\ -1 & 0 \end{bmatrix}, \quad E = \begin{bmatrix} 1 \\ 2 \\ 3 \end{bmatrix}$$

that are meaningful.

11 Verify the associative law of matrix multiplication by carrying out the multiplications in both ways.

$$\text{(a)} \quad \begin{bmatrix} 2 & 3 \\ 1 & 4 \end{bmatrix}\begin{bmatrix} 0 & 2 \\ 1 & 2 \end{bmatrix}\begin{bmatrix} 1 & -1 & 0 \\ 3 & 1 & 2 \end{bmatrix} \qquad \text{(b)} \quad [1 \ \ 2 \ \ 3]\begin{bmatrix} 2 & 0 & 1 \\ 0 & 4 & 3 \\ -1 & -2 & 1 \end{bmatrix}\begin{bmatrix} 5 \\ 1 \\ -5 \end{bmatrix}$$

$$\text{(c)} \quad \begin{bmatrix} 2 & -1 & 3 \\ 1 & 0 & 4 \\ -2 & -1 & 4 \end{bmatrix}\begin{bmatrix} 1 & 3 & 0 \\ 2 & 1 & 3 \\ 1 & 0 & 1 \end{bmatrix}\begin{bmatrix} 3 & 3 \\ 1 & 0 \\ -1 & 3 \end{bmatrix}$$

12 Let

$$A = \begin{bmatrix} 1 & 0 \\ 0 & 2 \\ 3 & 4 \end{bmatrix}, \qquad B = [-1 \ \ 2 \ \ 1], \qquad C = \begin{bmatrix} 3 & 1 \\ 0 & 1 \end{bmatrix}.$$

For which one of the six possible orders ABC, ACB, BAC, BCA, CAB, CBA is the product of these three matrices defined? Compute this product.

13 If a_{ij} denotes the entry in the ith row and the jth column of the $m \times n$ matrix A, and if b_{ij} denotes the entry in the ith row and the jth column of the $n \times k$ matrix B, then matrix multiplication is described by the formula $\sum_{r=1}^{n} a_{ir}b_{rj}$. Explain this formula. Show that the product of three matrices is given by the formula $\sum_{r=1}^{n} \sum_{s=1}^{k} a_{ir}b_{rs}c_{sj}$. This formula gives an alternate proof of the associativity of matrix multiplication.

Chapter 2

Equivalence of Matrices. Reduction to Diagonal Form

1 Tilts

CONSIDER the linear substitution

$$\begin{aligned} x &= a + b \\ y &= b \\ z &= c \end{aligned} \tag{1}$$

with matrix of coefficients $\begin{bmatrix} 1 & 1 & 0 \\ 0 & 1 & 0 \\ 0 & 0 & 1 \end{bmatrix}$. This substitution is easy to *invert.* That is, while the substitution expresses x, y, and z in terms of a, b, and c, one can easily express a, b, and c in terms of x, y, and z; merely subtract the second equation of (1) from the first to find

$$\begin{aligned} x - y &= a \\ y &= b \\ z &= c, \end{aligned}$$

or, in the format of a linear substitution,

$$\begin{aligned} a &= x - y \\ b &= y \\ c &= z. \end{aligned} \tag{2}$$

The substitution of (2) into (1) gives the equations $x = (x - y) + y = x$, $y = y$, $z = z$, or, in terms of a matrix product,

$$\begin{bmatrix} 1 & 1 & 0 \\ 0 & 1 & 0 \\ 0 & 0 & 1 \end{bmatrix}\begin{bmatrix} 1 & -1 & 0 \\ 0 & 1 & 0 \\ 0 & 0 & 1 \end{bmatrix} = \begin{bmatrix} 1 & 0 & 0 \\ 0 & 1 & 0 \\ 0 & 0 & 1 \end{bmatrix}. \tag{3}$$

Similarly, substitution of (1) into (2) gives $a = (a + b) - b = a$, $b = b$, $c = c$, a fact expressed by the matrix equation

$$\begin{bmatrix} 1 & -1 & 0 \\ 0 & 1 & 0 \\ 0 & 0 & 1 \end{bmatrix}\begin{bmatrix} 1 & 1 & 0 \\ 0 & 1 & 0 \\ 0 & 0 & 1 \end{bmatrix} = \begin{bmatrix} 1 & 0 & 0 \\ 0 & 1 & 0 \\ 0 & 0 & 1 \end{bmatrix}. \tag{4}$$

The matrix $\begin{bmatrix} 1 & 0 & 0 \\ 0 & 1 & 0 \\ 0 & 0 & 1 \end{bmatrix}$ on the right sides of (3) and (4) is the matrix of coefficients of a linear substitution in which the original variables and the new variables are identical and in the same order; it is called an *identity matrix.* Equations (3) and (4) say that the matrices

$$\begin{bmatrix} 1 & 1 & 0 \\ 0 & 1 & 0 \\ 0 & 0 & 1 \end{bmatrix} \quad \text{and} \quad \begin{bmatrix} 1 & -1 & 0 \\ 0 & 1 & 0 \\ 0 & 0 & 1 \end{bmatrix}$$

are *inverse* to each other—their product, in either order, is the identity matrix.

The **diagonal entries** of a matrix are the first entry of the first row, the second entry of the second row, the third entry of the third row, and so forth. In the case of square matrices, the diagonal entries truly lie on a diagonal of the square, the last one being the last entry of the last row. We will also use the term "diagonal entry," in the sense it is defined here, for matrices that are not square, even though it is not entirely apt in these cases.

An **identity matrix** is a square matrix whose diagonal entries are 1 and whose other entries are 0. The $n \times n$ identity matrix will be represented by I_n or, if its size is clear from the context, by I. If A has n rows, then, as is easily seen from the definition of matrix multiplication, $I_n A = A$. Similarly, if A has n columns, then $AI_n = A$.

By a **tilt*** we will mean a matrix such as the factors on the left side of (3) that is an identity matrix except that in one row an entry *next to* the diagonal is 1 or -1. Otherwise stated, a tilt is a matrix that is the matrix of coefficients of a linear substitution such as (1) or (2) in which the number of original variables is equal to the number of new variables, and in which the original variables and the new variables coincide, except that one new variable is equal to the corresponding original variable plus or minus either the preceding or the following original variable.

Like the two tilts above, any tilt is easy to invert. Loosely speaking, if a tilt adds a variable to an adjacent variable, the inverse tilt subtracts it from the adjacent variable; similarly, if a tilt subtracts a variable from an adjacent variable, the inverse tilt adds it back. In terms of matrices, the **inverse** of a tilt is represented by the same matrix, except that the sign of the one nonzero entry off the diagonal is reversed. The inverse of a tilt T will be denoted T^{-1}. With this notation, $TT^{-1} = I$ and $T^{-1}T = I$, as in equations (3) and (4).

---

*This is a made-up name. It is suggested by the geometry of the substitution $x = x + y$, $y = y$, which tilts the y-axis to an angle of 45°. A transformation of this general type is known in geometry as a shear, but the term "shear" is much more general, including shears to angles other than 45°.

2 Composition with Tilts

The rule for multiplication on the left by a tilt is easily inferred from an example, such as

$$\begin{bmatrix} 1 & 1 & 0 \\ 0 & 1 & 0 \\ 0 & 0 & 1 \end{bmatrix} \begin{bmatrix} 1 & 2 \\ 3 & 4 \\ 5 & 6 \end{bmatrix} = \begin{bmatrix} 4 & 6 \\ 3 & 4 \\ 5 & 6 \end{bmatrix}.$$

The second and third rows of the second factor on the left are unchanged in the product on the right. The first row is changed, but, as a review of the multiplication process shows, the new first row is simply the old first row plus the second row. In other words, multiplication of $\begin{bmatrix} 1 & 2 \\ 3 & 4 \\ 5 & 6 \end{bmatrix}$ on the left by $\begin{bmatrix} 1 & 1 & 0 \\ 0 & 1 & 0 \\ 0 & 0 & 1 \end{bmatrix}$ adds the second row to the first and leaves the second and third rows unchanged. Similarly,

$$\begin{bmatrix} 1 & 0 & 0 & 0 \\ 0 & 1 & 0 & 0 \\ 0 & 0 & 1 & -1 \\ 0 & 0 & 0 & 1 \end{bmatrix} \begin{bmatrix} 1 & 2 & 3 & 4 & 5 \\ 6 & 7 & 8 & 9 & 10 \\ 11 & 12 & 13 & 14 & 15 \\ 16 & 17 & 18 & 19 & 20 \end{bmatrix} = \begin{bmatrix} 1 & 2 & 3 & 4 & 5 \\ 6 & 7 & 8 & 9 & 10 \\ -5 & -5 & -5 & -5 & -5 \\ 16 & 17 & 18 & 19 & 20 \end{bmatrix}$$

shows that multiplication of this matrix on the left by this tilt has the effect of subtracting the fourth row from the third. Note that, in each of these cases, the tilt which is the first factor on the left is obtained by applying the operation in question—adding the second row to the first in the first case, subtracting the fourth row from the third in the second case—to the identity matrix with the same number of rows.

The general rule is simple: Let T be a given tilt, say an $n \times n$ tilt. It can be obtained from the $n \times n$ identity matrix I_n by adding a row to an adjacent row or by subtracting a row from an adjacent row. Multiplying any matrix M with n rows on the left by T gives the same result as applying to M the operation that transforms I_n into T.

Multiplication on the *right* by a tilt has a similar effect, the only difference being that a *column* is added to or subtracted from an adjacent one, rather than a row. For example, the tilt $\begin{bmatrix} 1 & 1 & 0 \\ 0 & 1 & 0 \\ 0 & 0 & 1 \end{bmatrix}$ in the first example above results from adding the first column of I_3 to the second. Therefore, multiplication on the right of a matrix M with 3 columns by this tilt has the effect of adding the first column of M to the second. For example,

$$\begin{bmatrix} 1 & 2 & 3 \\ 4 & 5 & 6 \end{bmatrix} \begin{bmatrix} 1 & 1 & 0 \\ 0 & 1 & 0 \\ 0 & 0 & 1 \end{bmatrix} = \begin{bmatrix} 1 & 3 & 3 \\ 4 & 9 & 6 \end{bmatrix}.$$

Similarly, the tilt $\begin{bmatrix} 1 & 0 & 0 & 0 \\ 0 & 1 & 0 & 0 \\ 0 & 0 & 1 & -1 \\ 0 & 0 & 0 & 1 \end{bmatrix}$ in the second example above is obtained by subtracting the third column from the fourth. Therefore, multiplying any matrix with 4 columns on the right by that tilt subtracts the third column from the fourth, as in the equation

$$[1 \quad 3 \quad 5 \quad 7] \begin{bmatrix} 1 & 0 & 0 & 0 \\ 0 & 1 & 0 & 0 \\ 0 & 0 & 1 & -1 \\ 0 & 0 & 0 & 1 \end{bmatrix} = [1 \quad 3 \quad 5 \quad 2].$$

3 *Equivalence of Matrices*

We will say that a matrix A is **linked** to a matrix B if $A = TB$ for some tilt T, or if $A = BT$ for some tilt T.

Since tilts are square, TB (or BT) has the same number of rows and columns as B, as follows immediately from the definition of matrix multiplication. Therefore, a matrix linked to A must be the same size as A.

As was shown in the preceding section, A is linked to B if and only if it can be obtained from B by adding a row to, or subtracting a row from, an adjacent row, or by adding a column to, or subtracting a column from, an adjacent column.

It was noted in Section 1 that every tilt T has an inverse T^{-1} which is also a tilt. Therefore, $A = TB$ implies $T^{-1}A = T^{-1}TB = IB = B$, where T^{-1} is a tilt. Similarly, if $A = BT$ then $AT^{-1} = B$. In other words, if A is linked to B then B is linked to A.

We will say that two matrices A and B are **equivalent** if $A = B$ or if there is a sequence of matrices $A_0, A_1, \ldots, A_k$, in which each A_i is linked to its predecessor A_{i-1} for $i = 1, 2, \ldots, k$, while $A_0 = A$ and $A_k = B$. Otherwise stated, A is equivalent to B if each can be transformed into the other by a sequence of operations (possibly *no* operations) in which a row or column is added to or subtracted from an adjacent one.

Note that this relation of equivalence has the three properties implied by the word "equivalent," namely, reflexivity, symmetry, and transitivity (respectively, any matrix is equivalent to itself, if A is equivalent to B then B is equivalent to A, and matrices equivalent to the same matrix are equivalent to each other).

In terms of tilts and matrix multiplication, this same definition can be phrased as follows: To say that A is equivalent to B means that B can be obtained from A by multiplying on the left and/or right by a succession of tilts. Thus, $B = MAN$, where M and N are products of tilts. (When the definition is stated in this way, identity matrices—including I_1—must be considered to be products of tilts. By convention, we will regard an identity matrix as a product of *no* tilts.* If A and B are $m \times n$ matrices then of course the tilts of which M is a product must be $m \times m$ and the tilts of which N is a product must be $n \times n$.)

There are no 1×1 tilts (no row has an adjacent row and no column has an adjacent column). Consequently, two 1×1 matrices are equivalent only if they are *equal*.

4 Unimodular Matrices

A matrix equivalent to an identity matrix is called **unimodular**. Thus, a unimodular matrix A is square, say $n \times n$, and satisfies $MI_nN = A$, where M and N are products of $n \times n$ tilts. A unimodular matrix is therefore a product of tilts. Since a product of $n \times n$ tilts is clearly equivalent to I_n, it follows that *a matrix is unimodular if and only if it is a product of tilts.* As was stated in Section 3, an identity matrix—even I_1—is regarded, by convention, as a product of tilts.

Therefore, the definition of equivalence can be restated: Two matrices A and B are equivalent if and only if $MAN = B$ where M and N are unimodular matrices (of the appropriate sizes).

As was noted in Section 1, every tilt T has an inverse T^{-1}, which is also a tilt. Therefore, if M is a unimodular matrix, say $M = T_1T_2 \cdots T_k$, then the matrix† $M^{-1} = T_k^{-1}T_{k-1}^{-1} \cdots T_1^{-1}$ is an *inverse* of M in the sense that both $M^{-1}M$ and MM^{-1} are I_n, as a simple computation shows. (For example, $MM^{-1} = T_1T_2 \cdots T_kT_k^{-1}T_{k-1}^{-1} \cdots T_1^{-1} = T_1T_2 \cdots T_{k-1}IT_{k-1}^{-1} \cdots T_1^{-1} = T_1T_2 \cdots T_{k-2}T_{k-2}^{-1} \cdots T_1^{-1} = \cdots = T_1T_1^{-1} = I$.)

By the invertibility of unimodular matrices, the equation $MAN = B$, where M and N are unimodular, implies $A = I_mAI_n = M^{-1}MANN^{-1} = M^{-1}BN^{-1}$, where M^{-1} and N^{-1} are unimodular. This was already noted, in effect, in Section 3 in the form of the statement that equivalence of matrices is a symmetric relation.

5 On Algorithms

A matrix is called **diagonal** if all entries other than the diagonal entries are zero. The objective of this chapter is to prove that every matrix is equivalent to a

*Although this convention may appear odd at first, it can be regarded as a natural extension of the convention that $a^0 = 1$ for any number a.

†Note that the inverse of a product (composition) is the product of the inverses *in the reverse order.* This is a familiar fact of everyday life. To invert the process of first putting on socks, then putting on shoes, one first removes shoes, then removes socks.

diagonal matrix. The proof will be by means of an *algorithm* for finding, given a matrix, an equivalent diagonal matrix.

The advent of computers has produced a fundamental change in mathematics by motivating mathematicians to *think algorithmically,* that is, to think in terms of specific computational procedures. At first, say in the 1950s and 1960s, computer science emphasized *efficient* algorithms, but, as computers grew in power, the usefulness of *simple* and *clear* algorithms—even those which are not particularly efficient—was increasingly recognized. If such algorithms are useful in computer science, where the computations are actually carried out, they are certainly useful in pure mathematics, where the computations are for the most part theoretical.

The algorithm of Section 7 for finding a diagonal matrix equivalent to a given matrix serves a purely theoretical purpose—it proves that every matrix is equivalent to a diagonal one. Therefore, its practicality is irrelevant. In this book the subject is not practical computation but rather the basic concepts of linear algebra. As in arithmetic, where the meaning of multiplication precedes the actual computation of products, the basic ideas come first.

6 The 1×2 Case

For 1×2 matrices, the statement that every matrix is equivalent to a diagonal one takes the form:

Theorem. *Every 1×2 matrix is equivalent to one in which the entry on the right is zero.*

The following algorithm generates a sequence of linked matrices, terminating with a matrix in which the entry on the right is zero.

Algorithm: As long as the entry on the right is not zero, perform the operation determined by:

1. If the two entries are equal, subtract the left entry from the right one.
2. If one entry is positive* and the other is not, add the positive entry to the other.
3. Otherwise, subtract the lesser† of the two entries from the greater.

---

*A number is positive if and only if it is greater than zero. In particular, zero is not positive.

†If neither number is positive, the "lesser" is the one with greater absolute value. For example, $-1 < 0$.

For example, if the given matrix is $[-24 \quad -9]$, the algorithm produces the sequence of linked matrices $[-24 \quad -9] \sim [-24 \quad 15] \sim [-9 \quad 15] \sim [6 \quad 15] \sim [6 \quad 9] \sim [6 \quad 3] \sim [3 \quad 3] \sim [3 \quad 0]$, where the symbol $\sim$ denotes the relation of equivalence of matrices. In this example, the second and third steps apply rule 2, the last step applies rule 1, and the others all apply rule 3. Another example is $[-7 \quad 24] \sim [17 \quad 24] \sim [17 \quad 7] \sim [10 \quad 7] \sim [3 \quad 7] \sim [3 \quad 4] \sim [3 \quad 1] \sim [2 \quad 1] \sim [1 \quad 1] \sim [1 \quad 0]$.

Proof that the algorithm terminates. Let $A = A_0$ be a given 1×2 matrix, and let $A_0 \sim A_1 \sim A_2 \sim \cdots$ be the sequence of 1×2 matrices generated by the above algorithm. It is to be shown that this sequence terminates. At each step $A_i \neq A_{i+1}$. Moreover, a negative entry of A_i cannot decrease in the step from A_i to A_{i+1} (when rule 1 or 3 applies and a negative entry changes, a negative number is subtracted from the entry that changes; when 2 applies, a positive one is added), and a positive entry cannot increase or become negative (it cannot be changed by rule 2, and it is decreased without becoming negative by rules 1 or 3). Finally, a positive entry cannot decrease to zero unless it is on the right, in which case it could never have been zero at an earlier stage. Therefore, once an entry changes, it can never return to its previous value. All the more so, no two matrices in the sequence are equal. Finally, if m is an integer greater than the absolute values of both entries of A then it is greater than the absolute values of both entries of A_1 and, for the same reason, greater than the absolute values of both entries of all later A_i. Therefore, the matrices A_i all lie in the set of $(2m + 1)^2$ matrices whose entries are in this range; since there are no repeats, the sequence must terminate, as was to be shown.

7 *The General Case*

Theorem. *Every matrix is equivalent to a diagonal matrix.*

This theorem will be proved by means of an algorithm that is an extension of the algorithm used to prove the 1×2 case. In the 1×2 case, there was just *one* nondiagonal entry to be reduced to zero, but in the general case there are many. The main new ingredient in the extended algorithm is an assignment of *priorities* to the nondiagonal entries, that is, a determination of which nondiagonal entries are to be reduced to zero first.

The highest priority will be assigned to the last entry of the first row. It will be followed by the entry to its left, then the entry to the left of that one, and so

forth, until the second entry of the first row is reached. The next priority after the second entry of the first row goes to the entry at the bottom of the first column, which is followed by the entry above it, then the entry above that one, until the next-to-the-top entry of the first column is reached. Once these nondiagonal entries have been reduced to zero, all entries in the first row and column, with the possible exception of the first diagonal entry, are zero. Next come the remaining nondiagonal entries of the second row and column in a similar sequence, from right to left in the second row and from bottom to top in the second column. Next come the remaining nondiagonal entries of the third row and column in the same order, and so forth. If the number of rows m in the matrix is the same as the number of columns n, the entry of lowest priority is the bottom entry of the next-to-last column. If $m > n$, the lowest priority goes to the entry of the last column just below the diagonal. If $m < n$, the lowest priority goes to the entry of the last row just to the right of the diagonal.

Algorithm: As long as the matrix is not diagonal, perform the operation determined as follows. Find the nonzero entry off the diagonal of highest priority. If it lies above the diagonal, determine the next operation as if the algorithm in the 1×2 case were being applied to the matrix whose entries are this entry and the entry to its left. If it lies below the diagonal, determine the next operation as if the algorithm in the 1×2 case were being applied to the matrix whose entries are this entry and the entry just above it.

Otherwise stated, if the nonzero entry off the diagonal of highest priority is the entry a_{ij} in the ith row of the jth column, the operation is determined as follows:

If $i < j$ the rules are:

1. If $a_{i,j-1} = a_{ij}$, subtract column $j - 1$ from column j.
2. If one of $a_{i,j-1}$ and a_{ij} is positive and the other is not, add the column which has the positive one to the column which has the other.
3. Otherwise, subtract the column containing the lesser of $a_{i,j-1}$ and a_{ij} from the column containing the other.

If $i > j$, the rules are:

1′. If $a_{i-1,j} = a_{ij}$, subtract row $i - 1$ from row i.

2′. If one of $a_{i-1,j}$ and a_{ij} is positive and the other is not, add the row which has the positive one to the row which has the other.

3′. Otherwise, subtract the row containing the lesser of $a_{i-1,j}$ and a_{ij} from the row containing the other.

For example, if the original matrix is $\begin{bmatrix} 2 & 4 & 6 \\ 1 & 3 & 5 \end{bmatrix}$, the algorithm generates the sequence of linked matrices

$$\begin{bmatrix} 2 & 4 & 6 \\ 1 & 3 & 5 \end{bmatrix} \sim \begin{bmatrix} 2 & 4 & 2 \\ 1 & 3 & 2 \end{bmatrix} \sim \begin{bmatrix} 2 & 2 & 2 \\ 1 & 1 & 2 \end{bmatrix} \sim \begin{bmatrix} 2 & 2 & 0 \\ 1 & 1 & 1 \end{bmatrix} \sim \begin{bmatrix} 2 & 0 & 0 \\ 1 & 0 & 1 \end{bmatrix} \sim$$
$$\sim \begin{bmatrix} 1 & 0 & -1 \\ 1 & 0 & 1 \end{bmatrix} \sim \begin{bmatrix} 1 & 1 & -1 \\ 1 & -1 & 1 \end{bmatrix} \sim \begin{bmatrix} 1 & 1 & 0 \\ 1 & -1 & 0 \end{bmatrix} \sim$$
$$\sim \begin{bmatrix} 1 & 0 & 0 \\ 1 & -2 & 0 \end{bmatrix} \sim \begin{bmatrix} 1 & 0 & 0 \\ 0 & -2 & 0 \end{bmatrix}$$

ending with a diagonal matrix. For another example,

$$\begin{bmatrix} 2 & 0 \\ 6 & 4 \\ 9 & 10 \end{bmatrix} \sim \begin{bmatrix} 2 & 0 \\ 6 & 4 \\ 3 & 6 \end{bmatrix} \sim \begin{bmatrix} 2 & 0 \\ 3 & -2 \\ 3 & 6 \end{bmatrix} \sim \begin{bmatrix} 2 & 0 \\ 3 & -2 \\ 0 & 8 \end{bmatrix} \sim \begin{bmatrix} 2 & 0 \\ 1 & -2 \\ 0 & 8 \end{bmatrix} \sim$$
$$\sim \begin{bmatrix} 1 & 2 \\ 1 & -2 \\ 0 & 8 \end{bmatrix} \sim \begin{bmatrix} 1 & 1 \\ 1 & -3 \\ 0 & 8 \end{bmatrix} \sim \begin{bmatrix} 1 & 0 \\ 1 & -4 \\ 0 & 8 \end{bmatrix} \sim \begin{bmatrix} 1 & 0 \\ 0 & -4 \\ 0 & 8 \end{bmatrix} \sim \begin{bmatrix} 1 & 0 \\ 0 & 4 \\ 0 & 8 \end{bmatrix} \sim$$
$$\sim \begin{bmatrix} 1 & 0 \\ 0 & 4 \\ 0 & 4 \end{bmatrix} \sim \begin{bmatrix} 1 & 0 \\ 0 & 4 \\ 0 & 0 \end{bmatrix}.$$

Note that, as occurs in both of these examples, all nondiagonal entries of the first row may be zero at some stage, only to have the reduction of the nondiagonal entries of the first column introduce nonzero entries into nondiagonal locations in the first row.

Proof that the algorithm terminates. If the given matrix has a nonzero entry in the first row to the right of the first entry, the first steps of the algorithm operate just like the algorithm of Section 6 on two adjacent entries of the first row, namely, the rightmost nonzero entry and its neighbor to the left. Therefore, after a finite number of steps, a matrix is reached in which what was the rightmost nonzero entry in the first row is reduced to zero. If there is still a nonzero entry to the right of the first entry then another finite number of steps reduces it to zero, and so forth, until a matrix is reached in which all entries in the first row, with the possible exception of the first, are zero.

If this matrix has a nonzero entry in the first column below the first entry, the succeeding steps of the algorithm operate just like the algorithm of Section 6 on two adjacent entries of the first column (namely, it operates on the nonzero entry farthest down the column and the entry immediately above it) *unless* the operation of this algorithm introduces nonzero entries into the first row by adding the second row to the first or subtracting the second row from the first. Since an operation of this type is performed only in cases in which it changes the first diagonal entry, a finite number of steps must either change the first diagonal entry or reach a matrix in which the first row and column contain only zeros in entries not on the diagonal.

If the first diagonal entry is zero or negative, the operation of the algorithm can only increase it; if it is positive, the operation of the algorithm can only decrease it while keeping it positive. Therefore, in the course of the algorithm it can only change a finite number of times. Therefore, what was just proved—that after a finite number of steps either the entries of the first row and column except for the first diagonal entry have all been reduced to zero or the first diagonal entry changes—implies that after a finite number of steps the entries of the first row and column other than the first diagonal entry must all be reduced to zero.

Once such a matrix is reached, the subsequent operations of the algorithm (if any) never change the first row and column because they add rows with zero in the first column to rows other than the first, or they subtract rows with zero in the first column from rows other than the first, or they do analogous operations on columns. Moreover, the operation of the algorithm on such a matrix is exactly like its operation on the matrix obtained by deleting the first row and column. Therefore, by what was just shown, after a finite number of additional steps the matrix must be reduced to one in which all entries of the first two rows and the first two columns, with the possible exception of the first two diagonal entries, are zero.

Continuing in this way, one must eventually reach a matrix in which all entries, with the possible exception of the diagonal entries, are zero, as was to be shown.

8 Finding Explicit Equivalences

The proof in the preceding section gives an algorithm for transforming a given matrix A into an equivalent diagonal matrix D, but it ignores the problem of finding unimodular matrices M and N for which $MAN = D$. All one needs to do to find M and N is to keep a record of the row and column additions or subtractions that were performed in transforming A to D. A simple way to keep such a record and to make the determination of M and N a part of the algorithm is the following device.

Given an $m \times n$ matrix A, construct an $(m+n) \times (m+n)$ matrix $\hat{A}$ in which A occupies the first m rows of the first n columns, in which the remaining columns of the first m rows contain I_m, in which the remaining rows of the first n columns contain I_n, and in which the lower right $n \times m$ corner contains only zeros. Schematically,

$$\hat{A} = \begin{bmatrix} A & I_m \\ I_n & 0 \end{bmatrix}.$$

Let the column and row additions and subtractions used to reduce A to D be performed on $\hat{A}$. When all these operations are completed, the result is a matrix of the form

$$\hat{D} = \begin{bmatrix} D & B \\ C & 0 \end{bmatrix},$$

that is, a matrix in which the upper left $m \times n$ corner is D and the lower right $n \times m$ corner has all entries zero. The $m \times m$ matrix B in the upper right corner is then M, and the $n \times n$ matrix C in the lower left corner is N, as can be seen as follows. Each time A is multiplied on the left by an $n \times n$ tilt T, $\hat{A}$ is multiplied on the left by an $(m+n) \times (m+n)$ tilt $\hat{T}$, namely, the tilt which is T in the upper left corner, I_m in the lower right corner, and zero elsewhere. Similarly, each time A is multiplied on the right by an $m \times m$ tilt T, $\hat{A}$ is multiplied on the right by the $(m+n) \times (m+n)$ tilt $\hat{T}$ which is T in the upper left corner, I_n in the lower right corner, and zero elsewhere. The product of all the tilts on the right is the matrix $\hat{N}$, which is N in the upper left corner, I_m in the lower right corner, and zero elsewhere. Similarly, the product of all the tilts on the left is $\hat{M}$. Thus, schematically,

$$\begin{bmatrix} M & 0 \\ 0 & I \end{bmatrix}\begin{bmatrix} A & I \\ I & 0 \end{bmatrix}\begin{bmatrix} N & 0 \\ 0 & I \end{bmatrix} = \begin{bmatrix} D & B \\ C & 0 \end{bmatrix}.$$

When the multiplications on the left are carried out, they give

$$\begin{bmatrix} MA & M \\ I & 0 \end{bmatrix}\begin{bmatrix} N & 0 \\ 0 & I \end{bmatrix} = \begin{bmatrix} MAN & M \\ N & 0 \end{bmatrix} = \begin{bmatrix} D & B \\ C & 0 \end{bmatrix},$$

which proves the desired conclusion $B = M$ and $C = N$.

Examples

If the given matrix is $\begin{bmatrix} 3 & 1 \\ 9 & 3 \end{bmatrix}$, the computation described in Section 8 takes the form

$$\begin{bmatrix} 3 & 1 & 1 & 0 \\ 9 & 3 & 0 & 1 \\ 1 & 0 & 0 & 0 \\ 0 & 1 & 0 & 0 \end{bmatrix} \sim \begin{bmatrix} 2 & 1 & 1 & 0 \\ 6 & 3 & 0 & 1 \\ 1 & 0 & 0 & 0 \\ -1 & 1 & 0 & 0 \end{bmatrix} \sim \begin{bmatrix} 1 & 1 & 1 & 0 \\ 3 & 3 & 0 & 1 \\ 1 & 0 & 0 & 0 \\ -2 & 1 & 0 & 0 \end{bmatrix} \sim$$

$$\sim \begin{bmatrix} 1 & 0 & 1 & 0 \\ 3 & 0 & 0 & 1 \\ 1 & -1 & 0 & 0 \\ -2 & 3 & 0 & 0 \end{bmatrix} \sim \begin{bmatrix} 1 & 0 & 1 & 0 \\ 2 & 0 & -1 & 1 \\ 1 & -1 & 0 & 0 \\ -2 & 3 & 0 & 0 \end{bmatrix} \sim$$

$$\sim \begin{bmatrix} 1 & 0 & 1 & 0 \\ 1 & 0 & -2 & 1 \\ 1 & -1 & 0 & 0 \\ -2 & 3 & 0 & 0 \end{bmatrix} \sim \begin{bmatrix} 1 & 0 & 1 & 0 \\ 0 & 0 & -3 & 1 \\ 1 & -1 & 0 & 0 \\ -2 & 3 & 0 & 0 \end{bmatrix},$$

leading to the explicit equivalence

$$\begin{bmatrix} 1 & 0 \\ -3 & 1 \end{bmatrix}\begin{bmatrix} 3 & 1 \\ 9 & 3 \end{bmatrix}\begin{bmatrix} 1 & -1 \\ -2 & 3 \end{bmatrix} = \begin{bmatrix} 1 & 0 \\ -3 & 1 \end{bmatrix}\begin{bmatrix} 1 & 0 \\ 3 & 0 \end{bmatrix} = \begin{bmatrix} 1 & 0 \\ 0 & 0 \end{bmatrix}$$

between the given matrix and a diagonal one.

If the given matrix is $\begin{bmatrix} 3 \\ -2 \\ 1 \end{bmatrix}$, the computation takes the form

$$\begin{bmatrix} 3 & 1 & 0 & 0 \\ -2 & 0 & 1 & 0 \\ 1 & 0 & 0 & 1 \\ 1 & 0 & 0 & 0 \end{bmatrix} \sim \begin{bmatrix} 3 & 1 & 0 & 0 \\ -1 & 0 & 1 & 1 \\ 1 & 0 & 0 & 1 \\ 1 & 0 & 0 & 0 \end{bmatrix} \sim \begin{bmatrix} 3 & 1 & 0 & 0 \\ 0 & 0 & 1 & 2 \\ 1 & 0 & 0 & 1 \\ 1 & 0 & 0 & 0 \end{bmatrix} \sim$$

$$\sim \begin{bmatrix} 3 & 1 & 0 & 0 \\ 1 & 0 & 1 & 3 \\ 1 & 0 & 0 & 1 \\ 1 & 0 & 0 & 0 \end{bmatrix} \sim \begin{bmatrix} 3 & 1 & 0 & 0 \\ 1 & 0 & 1 & 3 \\ 0 & 0 & -1 & -2 \\ 1 & 0 & 0 & 0 \end{bmatrix} \sim \begin{bmatrix} 2 & 1 & -1 & -3 \\ 1 & 0 & 1 & 3 \\ 0 & 0 & -1 & -2 \\ 1 & 0 & 0 & 0 \end{bmatrix} \sim$$

$$\sim \begin{bmatrix} 1 & 1 & -2 & -6 \\ 1 & 0 & 1 & 3 \\ 0 & 0 & -1 & -2 \\ 1 & 0 & 0 & 0 \end{bmatrix} \sim \begin{bmatrix} 1 & 1 & -2 & -6 \\ 0 & -1 & 3 & 9 \\ 0 & 0 & -1 & -2 \\ 1 & 0 & 0 & 0 \end{bmatrix}$$

leading to the explicit equivalence

$$\begin{bmatrix} 1 & -2 & -6 \\ -1 & 3 & 9 \\ 0 & -1 & -2 \end{bmatrix}\begin{bmatrix} 3 \\ -2 \\ 1 \end{bmatrix}[\,1\,] = \begin{bmatrix} 1 \\ 0 \\ 0 \end{bmatrix}$$

between the given matrix and a diagonal one. Note that the last row $[\,1 \quad 0 \quad 0 \quad 0\,]$ never changes during the course of the algorithm. The same will obviously be true whenever the given A has only one column; therefore, the last row of $\hat{A}$ can be left out when A has only one column. Similarly, the last column of $\hat{A}$ can be left out when A has only one row.

If the given matrix is $\begin{bmatrix} 0 & 2 & 1 \\ 2 & 4 & 2 \end{bmatrix}$, the computation takes the form

$$\begin{bmatrix} 0 & 2 & 1 & 1 & 0 \\ 2 & 4 & 2 & 0 & 1 \\ 1 & 0 & 0 & 0 & 0 \\ 0 & 1 & 0 & 0 & 0 \\ 0 & 0 & 1 & 0 & 0 \end{bmatrix} \sim \begin{bmatrix} 0 & 1 & 1 & 1 & 0 \\ 2 & 2 & 2 & 0 & 1 \\ 1 & 0 & 0 & 0 & 0 \\ 0 & 1 & 0 & 0 & 0 \\ 0 & -1 & 1 & 0 & 0 \end{bmatrix} \sim \begin{bmatrix} 0 & 1 & 0 & 1 & 0 \\ 2 & 2 & 0 & 0 & 1 \\ 1 & 0 & 0 & 0 & 0 \\ 0 & 1 & -1 & 0 & 0 \\ 0 & -1 & 2 & 0 & 0 \end{bmatrix} \sim$$

$$\sim \begin{bmatrix} 1 & 1 & 0 & 1 & 0 \\ 4 & 2 & 0 & 0 & 1 \\ 1 & 0 & 0 & 0 & 0 \\ 1 & 1 & -1 & 0 & 0 \\ -1 & -1 & 2 & 0 & 0 \end{bmatrix} \sim \begin{bmatrix} 1 & 0 & 0 & 1 & 0 \\ 4 & -2 & 0 & 0 & 1 \\ 1 & -1 & 0 & 0 & 0 \\ 1 & 0 & -1 & 0 & 0 \\ -1 & 0 & 2 & 0 & 0 \end{bmatrix} \sim$$

$$\sim \begin{bmatrix} 1 & 0 & 0 & 1 & 0 \\ 3 & -2 & 0 & -1 & 1 \\ 1 & -1 & 0 & 0 & 0 \\ 1 & 0 & -1 & 0 & 0 \\ -1 & 0 & 2 & 0 & 0 \end{bmatrix} \sim \begin{bmatrix} 1 & 0 & 0 & 1 & 0 \\ 2 & -2 & 0 & -2 & 1 \\ 1 & -1 & 0 & 0 & 0 \\ 1 & 0 & -1 & 0 & 0 \\ -1 & 0 & 2 & 0 & 0 \end{bmatrix} \sim$$

$$\sim \begin{bmatrix} 1 & 0 & 0 & 1 & 0 \\ 1 & -2 & 0 & -3 & 1 \\ 1 & -1 & 0 & 0 & 0 \\ 1 & 0 & -1 & 0 & 0 \\ -1 & 0 & 2 & 0 & 0 \end{bmatrix} \sim \begin{bmatrix} 1 & 0 & 0 & 1 & 0 \\ 0 & -2 & 0 & -4 & 1 \\ 1 & -1 & 0 & 0 & 0 \\ 1 & 0 & -1 & 0 & 0 \\ -1 & 0 & 2 & 0 & 0 \end{bmatrix}$$

leading to the explicit equivalence $\begin{bmatrix} 1 & 0 \\ -4 & 1 \end{bmatrix}\begin{bmatrix} 0 & 2 & 1 \\ 2 & 4 & 2 \end{bmatrix}\begin{bmatrix} 1 & -1 & 0 \\ 1 & 0 & -1 \\ -1 & 0 & 2 \end{bmatrix} =$ $\begin{bmatrix} 1 & 0 & 0 \\ 0 & -2 & 0 \end{bmatrix}$ between the given matrix and a diagonal one.

A diagonal matrix equivalent to $\begin{bmatrix} 0 & -2 & 0 & -2 \\ 1 & 4 & 0 & 6 \\ 2 & 4 & 6 & 8 \end{bmatrix}$ is found by

$$\begin{bmatrix} 0 & -2 & 0 & -2 \\ 1 & 4 & 0 & 6 \\ 2 & 4 & 6 & 8 \end{bmatrix} \sim \begin{bmatrix} 0 & -2 & 2 & -2 \\ 1 & 4 & -6 & 6 \\ 2 & 4 & -2 & 8 \end{bmatrix} \sim \begin{bmatrix} 0 & -2 & 2 & 0 \\ 1 & 4 & -6 & 0 \\ 2 & 4 & -2 & 6 \end{bmatrix} \sim$$

$$\sim \begin{bmatrix} 0 & 0 & 2 & 0 \\ 1 & -2 & -6 & 0 \\ 2 & 2 & -2 & 6 \end{bmatrix} \sim \begin{bmatrix} 0 & 2 & 2 & 0 \\ 1 & -8 & -6 & 0 \\ 2 & 0 & -2 & 6 \end{bmatrix} \sim \begin{bmatrix} 0 & 2 & 0 & 0 \\ 1 & -8 & 2 & 0 \\ 2 & 0 & -2 & 6 \end{bmatrix} \sim$$

$$\sim \begin{bmatrix} 2 & 2 & 0 & 0 \\ -7 & -8 & 2 & 0 \\ 2 & 0 & -2 & 6 \end{bmatrix} \sim \begin{bmatrix} 2 & 0 & 0 & 0 \\ -7 & -1 & 2 & 0 \\ 2 & -2 & -2 & 6 \end{bmatrix} \sim \begin{bmatrix} 2 & 0 & 0 & 0 \\ -5 & -3 & 0 & 6 \\ 2 & -2 & -2 & 6 \end{bmatrix} \sim$$

$$\sim \begin{bmatrix} 2 & 0 & 0 & 0 \\ -3 & -5 & -2 & 12 \\ 2 & -2 & -2 & 6 \end{bmatrix} \sim \begin{bmatrix} 2 & 0 & 0 & 0 \\ -1 & -7 & -4 & 18 \\ 2 & -2 & -2 & 6 \end{bmatrix} \sim \begin{bmatrix} 2 & 0 & 0 & 0 \\ 1 & -9 & -6 & 24 \\ 2 & -2 & -2 & 6 \end{bmatrix} \sim$$

$$\sim \begin{bmatrix} 2 & 0 & 0 & 0 \\ 1 & -9 & -6 & 24 \\ 1 & 7 & 4 & -18 \end{bmatrix} \sim \begin{bmatrix} 2 & 0 & 0 & 0 \\ 1 & -9 & -6 & 24 \\ 0 & 16 & 10 & -42 \end{bmatrix} \sim \begin{bmatrix} 1 & 9 & 6 & -24 \\ 1 & -9 & -6 & 24 \\ 0 & 16 & 10 & -42 \end{bmatrix} \sim$$

$$\sim \begin{bmatrix} 1 & 9 & 6 & -18 \\ 1 & -9 & -6 & 18 \\ 0 & 16 & 10 & -32 \end{bmatrix} \sim \begin{bmatrix} 1 & 9 & 6 & -12 \\ 1 & -9 & -6 & 12 \\ 0 & 16 & 10 & -22 \end{bmatrix} \sim \begin{bmatrix} 1 & 9 & 6 & -6 \\ 1 & -9 & -6 & 6 \\ 0 & 16 & 10 & -12 \end{bmatrix} \sim$$

$$\sim \begin{bmatrix} 1 & 9 & 6 & 0 \\ 1 & -9 & -6 & 0 \\ 0 & 16 & 10 & -2 \end{bmatrix} \sim \begin{bmatrix} 1 & 3 & 6 & 0 \\ 1 & -3 & -6 & 0 \\ 0 & 6 & 10 & -2 \end{bmatrix} \sim \begin{bmatrix} 1 & 3 & 3 & 0 \\ 1 & -3 & -3 & 0 \\ 0 & 6 & 4 & -2 \end{bmatrix} \sim$$

$$\sim \begin{bmatrix} 1 & 3 & 0 & 0 \\ 1 & -3 & 0 & 0 \\ 0 & 6 & -2 & -2 \end{bmatrix} \sim \begin{bmatrix} 1 & 2 & 0 & 0 \\ 1 & -4 & 0 & 0 \\ 0 & 6 & -2 & -2 \end{bmatrix} \sim \begin{bmatrix} 1 & 1 & 0 & 0 \\ 1 & -5 & 0 & 0 \\ 0 & 6 & -2 & -2 \end{bmatrix} \sim$$

$$\sim \begin{bmatrix} 1 & 0 & 0 & 0 \\ 1 & -6 & 0 & 0 \\ 0 & 6 & -2 & -2 \end{bmatrix} \sim \begin{bmatrix} 1 & 0 & 0 & 0 \\ 0 & -6 & 0 & 0 \\ 0 & 6 & -2 & -2 \end{bmatrix} \sim \begin{bmatrix} 1 & 0 & 0 & 0 \\ 0 & 0 & -2 & -2 \\ 0 & 6 & -2 & -2 \end{bmatrix} \sim$$

$$\sim \begin{bmatrix} 1 & 0 & 0 & 0 \\ 0 & 0 & -2 & 0 \\ 0 & 6 & -2 & 0 \end{bmatrix} \sim \begin{bmatrix} 1 & 0 & 0 & 0 \\ 0 & 2 & -2 & 0 \\ 0 & 8 & -2 & 0 \end{bmatrix} \sim \begin{bmatrix} 1 & 0 & 0 & 0 \\ 0 & 2 & 0 & 0 \\ 0 & 8 & 6 & 0 \end{bmatrix} \sim$$

$$\sim \begin{bmatrix} 1 & 0 & 0 & 0 \\ 0 & 2 & 0 & 0 \\ 0 & 6 & 6 & 0 \end{bmatrix} \sim \begin{bmatrix} 1 & 0 & 0 & 0 \\ 0 & 2 & 0 & 0 \\ 0 & 4 & 6 & 0 \end{bmatrix} \sim \begin{bmatrix} 1 & 0 & 0 & 0 \\ 0 & 2 & 0 & 0 \\ 0 & 2 & 6 & 0 \end{bmatrix} \sim \begin{bmatrix} 1 & 0 & 0 & 0 \\ 0 & 2 & 0 & 0 \\ 0 & 0 & 6 & 0 \end{bmatrix}.$$

Note that the problem is much easier if only D is required, not M and N.

Exercises

1 For each of the following matrices A, apply the method of Section 8 to find unimodular matrices M and N such that MAN is diagonal.

$$\text{(a)}\ \begin{bmatrix} 4 & 2 \\ -2 & -4 \end{bmatrix} \quad \text{(b)}\ \begin{bmatrix} 2 & 1 \\ 1 & 1 \\ 2 & -1 \end{bmatrix} \quad \text{(c)}\ \begin{bmatrix} 3 \\ 4 \\ 5 \end{bmatrix} \quad \text{(d)}\ \begin{bmatrix} 3 & 2 & 2 \\ 5 & 2 & 1 \end{bmatrix}$$

2 Compute the following matrix product in three ways: (a) Multiply successive pairs of factors in the product to write it as a product of four factors, multiply successive pairs in this product to write it as a product of two factors, and finally, multiply these two. (b) Use the fact that multiplication on the right by a tilt is a column operation to find the product first of the first two factors, then of the first three, the first four, and so forth, until finally the product of all eight is found. (c) Use the fact that multiplication on the left by a tilt is a row operation to find first the product of the last two factors, then the last three, and so forth.

$$\begin{bmatrix} 1 & 0 & 0 \\ 0 & 1 & 0 \\ 0 & 1 & 1 \end{bmatrix}\begin{bmatrix} 1 & -1 & 0 \\ 0 & 1 & 0 \\ 0 & 0 & 1 \end{bmatrix}\begin{bmatrix} 1 & 0 & 0 \\ -1 & 1 & 0 \\ 0 & 0 & 1 \end{bmatrix}\begin{bmatrix} 1 & 0 & 0 \\ 0 & 1 & 1 \\ 0 & 0 & 1 \end{bmatrix}\begin{bmatrix} 1 & 0 & 0 \\ 0 & 1 & 0 \\ 0 & 1 & 1 \end{bmatrix}\begin{bmatrix} 1 & -1 & 0 \\ 0 & 1 & 0 \\ 0 & 0 & 1 \end{bmatrix}\begin{bmatrix} 1 & 0 & 0 \\ -1 & 1 & 0 \\ 0 & 0 & 1 \end{bmatrix}\begin{bmatrix} 1 & 0 & 0 \\ 0 & 1 & 1 \\ 0 & 0 & 1 \end{bmatrix}$$

3 Show that half of the 2×2 tilts commute with

$$T = \begin{bmatrix} 1 & 0 \\ 1 & 1 \end{bmatrix}$$

and half do not. That is, half of the 2×2 tilts T' satisfy $TT' = T'T$. How many 3×3 tilts commute with

$$\begin{bmatrix} 1 & 0 & 0 \\ 1 & 1 & 0 \\ 0 & 0 & 1 \end{bmatrix}?$$

4 How many $n \times n$ tilts are there?

5 Find a general rule for determining whether two $n \times n$ tilts commute.

6 Prove that a 1×2 matrix $\begin{bmatrix} a & b \end{bmatrix}$ is equivalent to exactly one matrix of the form $\begin{bmatrix} d & 0 \end{bmatrix}$ in which $d \geq 0$.

7 The previous exercise shows that for any 1×2 matrix $\begin{bmatrix} a & b \end{bmatrix}$ there is a unique integer $d \geq 0$ such that $\begin{bmatrix} a & b \end{bmatrix} = \begin{bmatrix} d & 0 \end{bmatrix} N$ for some unimodular matrix N. Prove from first principles that when a and b are both *positive* there is such a unimodular matrix N which is a product in which the only factors are

$$R = \begin{bmatrix} 1 & 1 \\ 0 & 1 \end{bmatrix} \quad \text{or} \quad L = \begin{bmatrix} 1 & 0 \\ 1 & 1 \end{bmatrix}.$$

(When a and b are positive, the algorithm of Section 6 is simply the Euclidean algorithm—repeatedly subtract the lesser from the greater for as long as the numbers are unequal—for finding the greatest common divisor of two numbers. See Book VII of Euclid's *Elements*.)

8 Find representations in the form of the preceding exercise of each of the following 1×2 matrices. Check your answer by computing the product of Rs and Ls. (a) $\begin{bmatrix} 9 & 15 \end{bmatrix}$. (b) $\begin{bmatrix} 12 & 17 \end{bmatrix}$. (c) $\begin{bmatrix} 37 & 27 \end{bmatrix}$. (d) $\begin{bmatrix} 104 & 65 \end{bmatrix}$. (e) $\begin{bmatrix} 42 & 53 \end{bmatrix}$. (f) $\begin{bmatrix} 54 & 64 \end{bmatrix}$.

9 Show that if a unimodular matrix can be written as a product of the tilts R and L as in the previous two exercises, then this can be done in only one way.

10 Show that if $N = \begin{bmatrix} r & s \\ t & u \end{bmatrix}$ is a product of the tilts R and L then r, s, t, and u are nonnegative integers and $ru = st + 1$.

11 Let a and b be positive integers, and let d be their greatest common divisor, that is, let d be the unique positive integer such that $\begin{bmatrix} a & b \end{bmatrix} \sim \begin{bmatrix} d & 0 \end{bmatrix}$. Prove that there exist positive integers p and q such that $pa = qb + d$.

12 Find a multiple of 59 that is one greater than a multiple of 26.

13 Prove the matrix equivalences $\begin{bmatrix} 2 & 0 \\ 0 & 3 \end{bmatrix} \sim \begin{bmatrix} 1 & 0 \\ 0 & 6 \end{bmatrix}$ and $\begin{bmatrix} 6 & 0 \\ 0 & 7 \end{bmatrix} \sim \begin{bmatrix} 1 & 0 \\ 0 & 42 \end{bmatrix}$.

14 Prove that the matrices $\begin{bmatrix} 2 & 0 \\ 0 & 2 \end{bmatrix}$ and $\begin{bmatrix} 1 & 0 \\ 0 & 4 \end{bmatrix}$ are *not* equivalent.

15 Show that $\begin{bmatrix} -1 & 0 \\ 0 & -1 \end{bmatrix}$ is unimodular. Write this matrix as a product of tilts.

16 Show that a diagonal square matrix whose diagonal entries are all ± 1 is unimodular whenever the number of times -1 occurs on the diagonal is *even*. Show that in at least one case a diagonal matrix of this type is *not* unimodular when the number of times -1 occurs is odd.

17 Show that if A is a unimodular $n \times n$ matrix and if $MAN = I_n$ where M and N are unimodular matrices, then NM is a matrix inverse to A. Use this principle to find inverses of the following unimodular matrices:

$$\text{(a)} \begin{bmatrix} 9 & 7 \\ 14 & 11 \end{bmatrix} \qquad \text{(b)} \begin{bmatrix} 1 & 1 & 1 \\ 2 & 3 & 3 \\ 2 & 3 & 4 \end{bmatrix}$$

18 For $i = 2, 3, 4, 5$, let S_i be the matrix obtained by adding column 1 to column i of I_5. Show that each S_i is unimodular. Construct your argument in such a way that it generalizes to prove that *adding any column of I_n to any other gives a unimodular matrix,* and from that to prove that *if any row (column) of a matrix A is added to (subtracted from) another row (column), the resulting matrix is equivalent to A.*

19 Write a computer program which accepts as input an integer m, an integer n, and an $m \times n$ matrix of integers A, and gives as output unimodular matrices M and N and a diagonal matrix D such that $MAN = D$.

20 Application of the algorithm of Section 7 to $\begin{bmatrix} 1 & 1000 \end{bmatrix}$ calls for subtraction of the right entry from the left 1000 times. Revise the algorithm in such a way that instead of adding or subtracting one entry from the other it adds or subtracts an appropriate power of 2 times the entry, so that, in the case of $\begin{bmatrix} 1 & 1000 \end{bmatrix}$ it goes through the steps $\sim \begin{bmatrix} 1 & 488 \end{bmatrix} \sim \begin{bmatrix} 1 & 232 \end{bmatrix} \sim \begin{bmatrix} 1 & 104 \end{bmatrix} \sim \begin{bmatrix} 1 & 40 \end{bmatrix} \sim \begin{bmatrix} 1 & 8 \end{bmatrix} \sim \begin{bmatrix} 1 & 0 \end{bmatrix}$.

Chapter 3

Matrix Division

1 Division

THE PROBLEM OF DIVISION of numbers is the problem of finding, given a and y, all* solutions x of $ax = y$. The problem of matrix division is the problem of finding, given two matrices A and Y, all matrices X such that $AX = Y$.

Recall that for AX to be defined, the number of rows of X must equal the number of columns of A. When this is the case, the number of rows of AX is the number of rows of A, and the number of its columns is the number of columns of X. Therefore, $AX = Y$ can be true only if A and Y have the same number of rows, X and Y have the same number of columns, and the number of columns of A is the number of rows of X. The problem of division is therefore stated more completely by: Given an $m \times n$ matrix A and given a matrix Y with m rows, find all matrices X such that $AX = Y$. Necessarily, X has n rows and the same number of columns as Y.

Because matrix multiplication is not commutative, the problem of solving $AX = Y$ is quite distinct from the problem of solving $XA = Y$. The solution of $AX = Y$ is therefore called **division of Y on the left by A,** and the solution of $XA = Y$ is called **division of Y on the right** by A. (For division of Y on the right by A to be a meaningful problem, A and Y must have the same number of *columns*. When this is the case, any solution X must have the same number of rows as Y, and the number of its columns must be the number of rows of A.) This chapter will deal only with the problem of division on the left. Division on the right is easily solved by analogous methods, as is seen in the exercises.

Throughout this chapter and the next two, "number" will mean "integer." Therefore, the division problem $2x = 1$ will be regarded as having *no solution.* Similarly, the matrix division problem $AX = Y$ will be regarded as having no solution if there is no matrix X with *integer entries* such that $AX = Y$.

---

*If by "number" one means "integer," the solution of this problem is as follows. If $a \neq 0$, there is either one solution (in which case a is said to divide y) or there is none. If $a = 0$, there is either no solution (if $y \neq 0$) or any integer is a solution.

2 Division on the Left by a Diagonal Matrix

If A is diagonal, the problem of division on the left by A is very simple, because multiplication of X on the left by A merely multiplies each row of X by the corresponding diagonal entry of A. For example,

$$\begin{bmatrix} 2 & 0 & 0 & 0 & 0 \\ 0 & 0 & 0 & 0 & 0 \\ 0 & 0 & 3 & 0 & 0 \\ 0 & 0 & 0 & 5 & 0 \end{bmatrix} \begin{bmatrix} -3 & -2 \\ -1 & 0 \\ 1 & 2 \\ 3 & 4 \\ 5 & 6 \end{bmatrix} = \begin{bmatrix} -6 & -4 \\ 0 & 0 \\ 3 & 6 \\ 15 & 20 \end{bmatrix}.$$

Here the first row of X is multiplied by 2, the second row by 0, the third by 3, and the fourth by 5; the fifth row of X plays no role in AX because the fifth column of A is zero.

This observation makes it easy to determine all possible matrices X when AX is given. For example, in the above product, if the second factor on the left is hidden, one can still deduce from knowledge of A and AX that X must be a 5×2 matrix whose first row is $-3, -2$, whose third row is $1, 2$, and whose fourth row is $3, 4$. One cannot, on the other hand, deduce any information about the second row of X or the fifth row of X, because they are ignored in the computation of AX. In fact, for this A and for $Y = \begin{bmatrix} -6 & -4 \\ 0 & 0 \\ 3 & 6 \\ 15 & 20 \end{bmatrix}$, the solutions X of $AX = Y$ are the matrices of the form

$$X = \begin{bmatrix} -3 & -2 \\ a & b \\ 1 & 2 \\ 3 & 4 \\ c & d \end{bmatrix}$$

where a, b, c, and d are arbitrary numbers.

Note that for this A the entries of AX necessarily satisfy rather strong conditions, namely, that the second row of AX must be zero, that all entries in its first row must be even, that all entries in its third row must be divisible by 3, and that all entries in its fourth row must be divisible by 5. Unless Y satisfies these conditions, there is no solution X of $AX = Y$. For any 4×2 matrix Y satisfying these conditions, however, one can find a solution X of $AX = Y$ by taking the first row of X to be the first row of Y divided by 2, the third row of X to be the third row of Y divided by 3, the fourth row of X to be the fourth row of Y divided by 5, and the second and fifth rows of X to be anything at all.

The solution of the general problem of division on the left by a diagonal matrix A is the same:

When A is the diagonal $m \times n$ matrix with diagonal entries $a_1, a_2, \ldots, a_k$ (where k is the minimum of m and n) and when Y is a given matrix with m rows, the solutions X of $AX = Y$ are found as follows. There are no solutions unless all entries of the ith row of Y are multiples of a_i for $i = 1, 2, \dots, k$. If $n < m$, there are no solutions X unless all rows of Y past the nth are zero. However, if these conditions are all met, there are solutions X, and the most general solution is found by taking the ith row of X to be the ith row of Y divided by a_i when $a_i \neq 0$, and by choosing the remaining entries of X arbitrarily.

3 *Division on the Left in the General Case*

Consider now the problem of division on the left by a matrix A that is not diagonal. Such a problem can be reduced to division on the left by a diagonal matrix—and therefore can be solved as in the last section—in the following way.

Let the algorithm of Chapter 2 be applied to A to find unimodular matrices M and N such that MAN is diagonal, say $MAN = D$. If X is a solution of $AX = Y$, then the matrix Z defined by $Z = N^{-1}X$ satisfies $DZ = MANN^{-1}X = MAX = MY$. Thus, Z is one of the matrices found by dividing MY on the left by the diagonal matrix D. Conversely, if Z is any solution of the problem of dividing MY on the left by D then the matrix X defined by $X = NZ$ satisfies $AX = ANZ = M^{-1}MANZ = M^{-1}DZ = M^{-1}MY = Y$. In short, *the solutions X of $AX = Y$ correspond one-to-one to solutions Z of $DZ = MY$ via $X = NZ$ and $N^{-1}X = Z$.*

Thus, the problem $AX = Y$ can be solved—that is, *all* solutions X can be found—by using the method of the last section to find all solutions Z of $DZ = MY$ and, for each of these solutions Z, setting $X = NZ$. (Here A and Y are given, and the algorithm of Chapter 2 is used to find M, N, and D.)

For example, if $A = \begin{bmatrix} 1 & -1 \\ 2 & 0 \\ 1 & 3 \end{bmatrix}$ and $Y = \begin{bmatrix} 1 & 3 \\ 0 & 10 \\ -3 & 11 \end{bmatrix}$, then the algorithm of Chapter 2 gives $M = \begin{bmatrix} 1 & 0 & 0 \\ -2 & 1 & 0 \\ 3 & -2 & 1 \end{bmatrix}$, $N = \begin{bmatrix} 1 & 1 \\ 0 & 1 \end{bmatrix}$, and $D = MAN = \begin{bmatrix} 1 & 0 \\ 0 & 2 \\ 0 & 0 \end{bmatrix}$. Thus $MY = \begin{bmatrix} 1 & 0 & 0 \\ -2 & 1 & 0 \\ 3 & -2 & 1 \end{bmatrix}\begin{bmatrix} 1 & 3 \\ 0 & 10 \\ -3 & 11 \end{bmatrix} = \begin{bmatrix} 1 & 3 \\ -2 & 4 \\ 0 & 0 \end{bmatrix}$. By the method of the last section, the only solution of $DZ = MY$ is therefore $Z = \begin{bmatrix} 1 & 3 \\ -1 & 2 \end{bmatrix}$, so the only solution X of $AX = Y$ is

$$X = NZ = \begin{bmatrix} 1 & 1 \\ 0 & 1 \end{bmatrix}\begin{bmatrix} 1 & 3 \\ -1 & 2 \end{bmatrix} = \begin{bmatrix} 0 & 5 \\ -1 & 2 \end{bmatrix}.$$

If $A = \begin{bmatrix} 1 & 2 \\ 2 & 4 \end{bmatrix}$ and $Y = \begin{bmatrix} -1 \\ -2 \end{bmatrix}$, then $M = \begin{bmatrix} 1 & 0 \\ -2 & 1 \end{bmatrix}$, $N = \begin{bmatrix} 1 & -2 \\ 0 & 1 \end{bmatrix}$, and $D = \begin{bmatrix} 1 & 0 \\ 0 & 0 \end{bmatrix}$. Thus $MY = \begin{bmatrix} -1 \\ 0 \end{bmatrix}$. The most general solution Z of

$\begin{bmatrix} 1 & 0 \\ 0 & 0 \end{bmatrix} Z = \begin{bmatrix} -1 \\ 0 \end{bmatrix}$ is $Z = \begin{bmatrix} -1 \\ a \end{bmatrix}$, where a is an arbitrary integer. Thus, the most general solution of $AX = Y$ is

$$X = NZ = \begin{bmatrix} 1 & -2 \\ 0 & 1 \end{bmatrix} \begin{bmatrix} -1 \\ a \end{bmatrix} = \begin{bmatrix} -1 - 2a \\ a \end{bmatrix},$$

where a is an arbitrary integer.

As a third example, consider the case in which the given A is $\begin{bmatrix} 3 & 1 \\ 5 & -1 \\ 1 & 3 \end{bmatrix}$ but $Y = \begin{bmatrix} r \\ s \\ t \end{bmatrix}$ is a matrix whose integer entries r, s, and t are to be given later. Then $M = \begin{bmatrix} 1 & 0 & 0 \\ -1 & 3 & 4 \\ 2 & -1 & -1 \end{bmatrix}$, $N = \begin{bmatrix} 1 & -1 \\ -2 & 3 \end{bmatrix}$, and $D = MAN = \begin{bmatrix} 1 & 0 \\ 0 & 8 \\ 0 & 0 \end{bmatrix}$, so the equation $DZ = MY$ is $\begin{bmatrix} 1 & 0 \\ 0 & 8 \\ 0 & 0 \end{bmatrix} Z = \begin{bmatrix} r \\ -r + 3s + 4t \\ 2r - s - t \end{bmatrix}$. Thus, there will be *no* solution of $AX = Y$ unless Y satisfies two conditions: $2r - s - t$ must be zero, and $-r + 3s + 4t$ must be divisible by 8. When these conditions are satisfied, there is a unique solution X, namely, $X = NZ$ where Z is the 2×1 matrix whose first entry is r and whose second entry is $(-r + 3s + 4t)/8$.

4 *Matrix Addition*

The relation between *one* solution X of $AX = Y$ and the *most general* solution is conveniently described in terms of *matrix addition.* If A and B are matrices of the same size, their **sum**, denoted $A + B$, is defined to be the matrix of the same size as A and B whose entry in each position is the sum of the entry of A in that position and the entry of B in that position. Similarly, the **difference** $A - B$ of two matrices of the same size is defined to be the matrix of the same size as A and B whose entry in each position is the entry of A in that position minus the entry of B in that position. (If A and B are not of the same size, then $A + B$ and $A - B$ have no meaning.)

As is easily seen from the definition, matrix multiplication is *distributive* in the usual way over matrix addition and subtraction. That is, if C has the same number of rows as A and B have columns, then $(A \pm B)C = AC \pm BC$, and if E has the same number of columns as A and B have rows, then $E(A \pm B) = EA \pm EB$.

If X_1 and X_2 are two solutions of $AX = Y$, then the matrix $U = X_1 - X_2$ satisfies* $AU = AX_1 - AX_2 = Y - Y = 0$. Conversely, if $AX_1 = Y$ and $AU = 0$

*Notation: If A is a matrix, then the equation $A = 0$ means that all entries of the matrix A are zero. Accordingly, $A \neq 0$ means that at least one entry of A is nonzero.

then the matrix $X_2 = X_1 + U$ satisfies $AX_2 = AX_1 + AU = AX_1 = Y$. Thus, *if X_1 is one solution of $AX = Y$ then the most general solution of $AX = Y$ is $X_1 + U$, where U is a matrix of the same size as X for which $AU = 0$.* That is, every solution is of the form $X_1 + U$, and everything of the form $X_1 + U$ is a solution.

In the case of a diagonal matrix A, $AU = 0$ if and only if U is zero in all rows corresponding to nonzero columns of A. Therefore, if X_1 is *one* solution of $AX = Y$, then the *most general* solution is $X_1 + U$, where U is a matrix of the same size as X_1 that is zero in rows corresponding to nonzero columns of A. This is just another way of saying that the entries of a solution are arbitrary in rows where A is zero and in rows past the mth row when $m < n$.

For any A, the most general solution X of $AX = Y$ is found by adding to any one solution X_1 the most general matrix U of the same size as X_1 for which $AU = 0$. If M and N are unimodular matrices such that $MAN = D$ is diagonal, the most general solution U of $AU = 0$ is $U = NV$, where V is any matrix of the right size for which $DV = 0$ (because $Y = 0$ implies $MY = 0$). As was just seen, the most general solution V of $DV = 0$ is a matrix which is zero in rows where D has nonzero entries. Thus, the most general solution X of $AX = Y$ is found by adding to any one solution X_1 a matrix of the form NV, where V is a matrix of the same size as X_1 that is zero in rows where D is nonzero.

5 *Zero Divisors*

For any matrix A, the equation $AX = 0$ has solutions $X = 0$, where X has the same number of rows as A has columns, and where X may have any number of columns. If there is a solution X of $AX = 0$ in which $X \neq 0$, then A is called a **left zero divisor**. Similarly, if $XA = 0$ has a solution $X \neq 0$, then A is a **right zero divisor**.

Let A be a given $m \times n$ matrix, and let D be a diagonal matrix equivalent to A. Clearly, A is a left zero divisor if and only if D is, because the solutions X of $AX = 0$ correspond one-to-one with solutions Z of $DZ = 0$ via $Z = N^{-1}X$, $X = NZ$ (because $MY = 0$ when $Y = 0$). But, as is obvious from the solution of Section 2, a diagonal matrix is a left zero divisor if and only if one of its columns is entirely zero.

Similarly, D, and therefore A, is a right zero divisor if and only if one of the rows of D is entirely zero. (See Exercise 13.)

We will say that an $m \times n$ matrix A is **high** if $n < m$, and **wide** if $n > m$. A matrix which is neither high nor wide is **square**.

Proposition 1. *A wide matrix is a left zero divisor. A high matrix is a right zero divisor.*

Proof. A wide diagonal matrix obviously has a column of zeros, and a high diagonal matrix obviously has a row of zeros.

Proposition 2. *A square matrix which has any one of the following properties has them all:*

(1) It is equivalent to a diagonal matrix with a diagonal entry equal to zero.

(2) It is a left zero divisor.

(3) It is a right zero divisor.

(4) Every diagonal matrix equivalent to it has a diagonal entry equal to zero.

Proof. Let D be a diagonal matrix equivalent to A. Since D is square, it has a column of zeros if and only if one of its diagonal entries is zero. Therefore, (1) implies (2) and (2) implies (1). In the same way, (1) implies (3) and (3) implies (1), because a square diagonal matrix has a row of zeros if and only if one of its diagonal entries is zero. Of course (4) implies (1). Conversely, (1) implies (4) because (1) implies (2), which implies (4). Therefore, each of the four conditions implies (1), and (1) implies each of the others.

A square matrix with these properties is called a **zero divisor**.

Examples

Let $A = \begin{bmatrix} 1 & 0 & 0 \\ 0 & 3 & 0 \end{bmatrix}$. For $Y = \begin{bmatrix} 1 & 2 \\ 3 & 4 \end{bmatrix}$, $AX = Y$ has no solution. For $Y = \begin{bmatrix} 1 & 2 \\ -3 & 9 \end{bmatrix}$ it has the solutions $X = \begin{bmatrix} 1 & 2 \\ -1 & 3 \\ a & b \end{bmatrix}$, where a and b are arbitrary integers.

Let $A = \begin{bmatrix} 1 & 0 \\ 0 & 3 \\ 0 & 0 \end{bmatrix}$. For $Y = \begin{bmatrix} 1 \\ 3 \\ 5 \end{bmatrix}$, $AX = Y$ has no solution. For $Y = \begin{bmatrix} 2 \\ 6 \\ 0 \end{bmatrix}$ the only solution is $X = \begin{bmatrix} 2 \\ 2 \end{bmatrix}$.

Let $A = \begin{bmatrix} 1 & 2 \\ 2 & 4 \end{bmatrix}$. Then $MAN = \begin{bmatrix} 1 & 0 \\ 0 & 0 \end{bmatrix}$ where $M = \begin{bmatrix} 1 & 0 \\ -2 & 1 \end{bmatrix}$ and $N = \begin{bmatrix} 1 & -2 \\ 0 & 1 \end{bmatrix}$. If $Y = \begin{bmatrix} 1 & 2 \\ 3 & 4 \end{bmatrix}$ then $MY = \begin{bmatrix} 1 & 2 \\ 1 & 0 \end{bmatrix}$; since the second row is not zero, $AX = Y$ has no solution. If $Y = \begin{bmatrix} 3 \\ 6 \end{bmatrix}$ then $MY = \begin{bmatrix} 3 \\ 0 \end{bmatrix}$. A solution of $AX = Y$ is $X = NZ$ where $Z = \begin{bmatrix} 3 \\ 0 \end{bmatrix}$, that is, $X = \begin{bmatrix} 3 \\ 0 \end{bmatrix}$. The most general solution is $X = \begin{bmatrix} 3 \\ 0 \end{bmatrix} + \begin{bmatrix} 1 & -2 \\ 0 & 1 \end{bmatrix}\begin{bmatrix} 0 \\ a \end{bmatrix} = \begin{bmatrix} 3-2a \\ a \end{bmatrix}$, where a is an arbitrary integer.

Let $A = \begin{bmatrix} 2 & 0 & 5 \\ 0 & 1 & 0 \\ 3 & 0 & 7 \end{bmatrix}$. The algorithm of Chapter 2 gives $MAN = \begin{bmatrix} 1 & 0 & 0 \\ 0 & 1 & 0 \\ 0 & 0 & -1 \end{bmatrix}$ where $M = \begin{bmatrix} 1 & 0 & 0 \\ -1 & 1 & 1 \\ -3 & 1 & 2 \end{bmatrix}$ and $N = \begin{bmatrix} 3 & -5 & -5 \\ -1 & 2 & 1 \\ -1 & 2 & 2 \end{bmatrix}$. There is a solution of $AX = Y$ for any matrix Y with three rows. It is found by multiplying the last row of MY by -1 and then multiplying the resulting matrix on the left by N. In other words, it is found by multiplying Y by $\begin{bmatrix} 3 & -5 & -5 \\ -1 & 2 & 1 \\ -1 & 2 & 2 \end{bmatrix}\begin{bmatrix} 1 & 0 & 0 \\ 0 & 1 & 0 \\ 0 & 0 & -1 \end{bmatrix}\begin{bmatrix} 1 & 0 & 0 \\ -1 & 1 & 1 \\ -3 & 1 & 2 \end{bmatrix} = \begin{bmatrix} 3 & -5 & -5 \\ -1 & 2 & 1 \\ -1 & 2 & 2 \end{bmatrix}\begin{bmatrix} 1 & 0 & 0 \\ -1 & 1 & 1 \\ 3 & -1 & -2 \end{bmatrix}$ $= \begin{bmatrix} -7 & 0 & 5 \\ 0 & 1 & 0 \\ 3 & 0 & -2 \end{bmatrix}$. This last matrix can be denoted A^{-1} because, as has been shown, $AX = Y$ if and only if $X = A^{-1}Y$, where A^{-1} is this matrix.

When $A = \begin{bmatrix} 3 & 1 & 2 \\ 2 & 4 & 6 \end{bmatrix}$ one finds

$$\begin{bmatrix} 1 & 0 \\ 6 & 1 \end{bmatrix}\begin{bmatrix} 3 & 1 & 2 \\ 2 & 4 & 6 \end{bmatrix}\begin{bmatrix} 1 & 0 & -1 \\ -2 & 2 & -7 \\ 0 & -1 & 5 \end{bmatrix} = \begin{bmatrix} 1 & 0 & 0 \\ 0 & 2 & 0 \end{bmatrix}.$$

The most general solution of $AX = \begin{bmatrix} 3 \\ 2 \end{bmatrix}$ is

$$X = \begin{bmatrix} 1 & 0 & -1 \\ -2 & 2 & -7 \\ 0 & -1 & 5 \end{bmatrix}\begin{bmatrix} 3 \\ 10 \\ a \end{bmatrix} = \begin{bmatrix} 3-a \\ 14-7a \\ -10+5a \end{bmatrix}.$$

Exercises

1 Find all matrices X for which $AX = Y$, where $A = \begin{bmatrix} 1 & 0 & 0 \\ 0 & 2 & 0 \end{bmatrix}$ and $Y = \begin{bmatrix} 4 \\ 4 \end{bmatrix}$.

2 Reformulate the problem of finding all triples of integers (x, y, z) satisfying

$$3x + y + 2z = 0$$

$$x + y - z = 0$$

as a problem in matrix division, and solve it.

3 Find all solutions X of $AX = Y$ where $A = \begin{bmatrix} 2 & 1 \\ -1 & 1 \end{bmatrix}$ and $Y = \begin{bmatrix} 9 & 8 \\ -3 & -1 \end{bmatrix}$.

4 A plane in xyz-space is described by an equation of the form $ax + by + cz + d = 0$. Describe the problem of finding the plane through the three points $(1, 2, 0)$, $(0, 2, 1)$, and $(1, 1, 1)$ as a problem in matrix division and solve it using the methods of this chapter. Then solve it by high-school algebra.

5 What conditions, if any, must a 3×1 matrix $Y = \begin{bmatrix} r \\ s \\ t \end{bmatrix}$ satisfy in order for there to be a solution X of $AX = Y$ when $A = \begin{bmatrix} 2 & 3 \\ 2 & 4 \\ 4 & 7 \end{bmatrix}$?

6 Find all solutions X of $AX = I$, where $A = \begin{bmatrix} 2 & 5 \\ 1 & 2 \end{bmatrix}$ and I is the 2×2 identity matrix.

7 A solution of $3x + 2y = 9$ is a matrix $X = \begin{bmatrix} x \\ y \end{bmatrix}$ satisfying $AX = Y$, where $A = \begin{bmatrix} 3 & 2 \end{bmatrix}$ and Y is the 1×1 matrix whose sole entry is 9. Find a solution by inspection. Find the most general solution of this division problem.

8 Give the most general solution of $x + 2y + 3z = 0$ by inspection and verify that the solution given by the method of the text agrees with it.

9 Find all solutions X of $AX = Y$ in the following three cases: (a) $A = \begin{bmatrix} 6 & 8 \end{bmatrix}$ and $Y = \begin{bmatrix} 2 & 0 \end{bmatrix}$. (b) $A = \begin{bmatrix} 6 \\ 8 \end{bmatrix}$ and $Y = \begin{bmatrix} 2 \\ 0 \end{bmatrix}$. (c) $A = \begin{bmatrix} 2 & 3 \\ 1 & 2 \end{bmatrix}$ and $Y = \begin{bmatrix} 1 & 0 \\ 0 & 1 \end{bmatrix}$.

10 For what integers w does the equation $105x + 57y = w$ have a solution x, y in integers? Give a formula for the solution that is valid whenever w satisfies your condition.

11 The matrix $A = \begin{bmatrix} 4 & 1 & -3 \\ 7 & 3 & -5 \end{bmatrix}$, being wide, must be a left zero divisor. Find a nonzero matrix X with one column such that $AX = 0$. What is the most general matrix with one column which satisfies this equation?

12 Give the solution of the problem of division on the right.

13 Let a given matrix A be equivalent to a diagonal matrix D. Show that A is a right zero divisor if and only if D contains a zero row.

14 A **right inverse** of an $m \times n$ matrix A is a solution B of $AB = I$, where necessarily $I = I_m$ and B is $n \times m$. Show that if A and A' are equivalent then A has a right inverse if and only if A' does. Describe explicitly the set of all diagonal matrices that have right inverses. Conclude that no high matrix has a right inverse.

15 A **left inverse** of a matrix A is a solution B of $BA = I$. Show that a wide matrix has no left inverse.

16 A matrix A is **invertible** if it has both a right inverse and a left inverse. The preceding two exercises imply that an invertible matrix must be square. Show that if A is invertible then it has only one right inverse and only one left inverse, and that these two are the same. Describe explicitly the set of all invertible diagonal matrices, and give a formula for the inverse of an invertible diagonal matrix. Prove that a matrix equivalent to an invertible matrix is invertible.

17 Prove that if A is invertible then either A is unimodular or the matrix obtained by reversing the signs in all entries of the last row of A is unimodular.

18 Let $A = \begin{bmatrix} 2 & 1 \\ 1 & 1 \\ 3 & 2 \end{bmatrix}$. For which 3×1 matrices Y does the equation $AX = Y$ have a solution X in integers? Answer the same question for $A = \begin{bmatrix} 1 & -1 \\ 2 & 0 \\ 3 & 1 \end{bmatrix}$.

Chapter 4

Determinants

1 Introduction

A **determinant** OF A SQUARE MATRIX is any number that is the product of the diagonal entries of a diagonal matrix equivalent to it. The main objective of this chapter is to show that *a square matrix has only one determinant;* in other words, if two diagonal square matrices are equivalent, then the product of the diagonal entries of one is the same as the product of the diagonal entries of the other. Once this theorem is proved, we will speak of *the* determinant of a square matrix, not *a* determinant.

In the case of 1×1 matrices, it is obvious that a matrix has only one determinant, because two 1×1 matrices are equivalent only if they are equal.

2 The 2×2 Case

Theorem. *The only determinant of a 2×2 matrix $\begin{bmatrix} a & b \\ c & d \end{bmatrix}$ is the number $ad - bc$.*

Notation: Given a 2×2 matrix $A = \begin{bmatrix} a & b \\ c & d \end{bmatrix}$, the number $ad - bc$ will be denoted $\begin{vmatrix} a & b \\ c & d \end{vmatrix}$ or $|A|$. Once the theorem is proved, this number can justly be called the determinant of A.

Proof. If

$$\begin{bmatrix} a & b \\ c & d \end{bmatrix}\begin{bmatrix} 1 & 1 \\ 0 & 1 \end{bmatrix} = \begin{bmatrix} a' & b' \\ c' & d' \end{bmatrix}$$

then $a' = a, b' = a+b, c' = c, d' = c+d$, so $a'd' - b'c' = a(c+d) - (a+b)c = ad - bc$. In other words, multiplication of a 2×2 matrix A on the right by the tilt $T = \begin{bmatrix} 1 & 1 \\ 0 & 1 \end{bmatrix}$ gives a matrix AT for which $|AT| = |A|$. Similar calculations show that $|AT| = |A|$ for *any* 2×2 tilt T and that $|TA| = |A|$ as well.

By the definition of equivalence, if $A = \begin{bmatrix} a & b \\ c & d \end{bmatrix}$ is equivalent to $B = \begin{bmatrix} a'' & b'' \\ c'' & d'' \end{bmatrix}$ then A can be transformed into B by a sequence of operations in which a matrix is multiplied on the left or right by a tilt. Each such operation leaves $ad - bc$ unchanged, so $ad - bc = a''d'' - b''c''$. If B is diagonal, then $a''d'' - b''c'' = a''d''$ is the product of the diagonal entries of B. Therefore, $ad - bc$ is equal to the product of the diagonal entries of any diagonal matrix equivalent to A, as was to be shown.

3 The 3×3 Case

The method of proof in the case of 3×3 matrices is the same: one gives a *formula* for the determinant.

Definition. Given a 3×3 matrix A, the number $|A|$ is defined by the formula

$$\begin{vmatrix} a_1 & a_2 & a_3 \\ b_1 & b_2 & b_3 \\ c_1 & c_2 & c_3 \end{vmatrix} = a_1b_2c_3 + a_2b_3c_1 + a_3b_1c_2 - a_3b_2c_1 - a_2b_1c_3 - a_1b_3c_2.$$

Theorem. *The only determinant of a* 3×3 *matrix* A *is the number* $|A|$.

Again, once the theorem is proved the number $|A|$ can justly be called the determinant of A.

Proof. As follows directly from the definition of $|A|$, if A is diagonal then $|A|$ is the product of the diagonal entries of A. What is to be shown, therefore, as in the 2×2 case, is that for any 3×3 matrix A and for any 3×3 tilt T, the formulas $|AT| = |A|$ and $|TA| = |A|$ hold. In the 2×2 case, there are 4 tilts, so there are 8 cases to be checked. In the 3×3 case there are 8 tilts and 16 cases to be checked. Naturally, the best way to prove the needed formulas is to discover the underlying reason they hold, not to write out the calculation in all 16 cases.

The formula for $|A|$ is a sum of 6 terms, and each term is plus or minus a product of three entries of A; moreover, in each term, the three entries that are multiplied together include exactly one entry from each row (one a, one b, and one c) and exactly one entry from each column (one subscript 1, one subscript 2, and one subscript 3); finally, each of the 6 ways of choosing one entry from each row and one entry in each column is used in exactly one term of the formula. These observations describe the formula for $|A|$ completely except for the crucial point: the sign in front of each term.

Multiplication of A on the right by the tilt $T = \begin{bmatrix} 1 & 0 & 0 \\ 0 & 1 & 0 \\ 0 & 1 & 1 \end{bmatrix}$ gives $\begin{bmatrix} a_1 & a_2 + a_3 & a_3 \\ b_1 & b_2 + b_3 & b_3 \\ c_1 & c_2 + c_3 & c_3 \end{bmatrix}$. Application of the formula to this matrix AT gives a sum of 6 terms, each of which splits into two, namely, $|AT| = a_1(b_2 + b_3)c_3 +$ (five more terms) $= a_1b_2c_3 + a_1b_3c_3 +$ (10 more terms). The 12 terms in this expression of $|AT|$ can be rearranged into two groups of six so that the formula takes the form

$$\begin{vmatrix} a_1 & a_2 + a_3 & a_3 \\ b_1 & b_2 + b_3 & b_3 \\ c_1 & c_2 + c_3 & c_3 \end{vmatrix} = \begin{vmatrix} a_1 & a_2 & a_3 \\ b_1 & b_2 & b_3 \\ c_1 & c_2 & c_3 \end{vmatrix} + \begin{vmatrix} a_1 & a_3 & a_3 \\ b_1 & b_3 & b_3 \\ c_1 & c_3 & c_3 \end{vmatrix}, \tag{1}$$

that is, $|AT| = |A| + |B|$, where B is the 3×3 matrix obtained by replacing the second column of A with the third column of A. (Note that the sum of the determinant of A and the determinant of B bears no relation to the determinant of the matrix $A + B$.) To prove that $|AT| = |A|$, it is necessary and sufficient, therefore, to prove that $|B| = 0$. But this is clear from the formula, which gives $|B| = a_1b_3c_3 + a_3b_3c_1 + a_3b_1c_3 - a_3b_3c_1 - a_3b_1c_3 - a_1b_3c_3$ (change each subscript 2 to a subscript 3) $= 0$ (each term occurs once with the sign $+$ and once with the sign $-$).

Similarly, if T is the tilt $\begin{bmatrix} 1 & -1 & 0 \\ 0 & 1 & 0 \\ 0 & 0 & 1 \end{bmatrix}$, one easily finds that $|TA| = |A| + |C|$ where

$$C = \begin{bmatrix} -b_1 & -b_2 & -b_3 \\ b_1 & b_2 & b_3 \\ c_1 & c_2 & c_3 \end{bmatrix}.$$

Directly from the formula, one finds that $|C| = (-b_1)b_2c_3 + (-b_2)b_3c_1 + (-b_3)b_1c_2 - (-b_3)b_2c_1 - (-b_2)b_1c_3 - (-b_1)b_3c_2 = 0$, so $|TA| = |A|$ in this case too.

As these calculations show, the essential property of the formula for $|A|$ that is needed to prove $|AT| = |A|$ and $|TA| = |A|$ is that *if B is a 3×3 matrix in which two adjacent rows are equal, in which two adjacent columns are equal, in which a row is equal to the negative of an adjacent row, or in which a column is equal to the negative of an adjacent column, then* $|B| = 0$. Since it is clear that reversing all signs in a row or column of a 3×3 matrix B reverses the sign of $|B|$, it suffices to prove that $|B| = 0$ in just the first two of these four cases, that is, that $|B| = 0$ whenever two adjacent rows or two adjacent columns of B are equal. This reduces the verifications to just four cases: when the first two columns of B are equal, when the last two columns of B are equal, when the first two rows of B are equal, and when the last two rows of B are equal. These four cases are easily checked, and the theorem follows.

(Note that the same argument reduces the 2×2 case to the verification of two formulas $\begin{vmatrix} a & b \\ a & b \end{vmatrix} = ab - ba = 0$ and $\begin{vmatrix} a & a \\ c & c \end{vmatrix} = ac - ac = 0$.)

Corollary. *The matrices* $\begin{bmatrix} 1 & 0 & 0 \\ 0 & 1 & 0 \\ 0 & 0 & 1 \end{bmatrix}$ *and* $\begin{bmatrix} 1 & 0 & 0 \\ 0 & 1 & 0 \\ 0 & 0 & -1 \end{bmatrix}$ *are not equivalent.*

(Equivalent matrices must have the same determinant.)

4 *The 4 × 4 Case*

The 2×2 and 3×3 cases lead one to expect that the 4×4 case will require a successful guess of a *formula* for the determinant $|A|$ of a 4×4 matrix A. Moreover, one expects that the formula for

$$\begin{vmatrix} a_1 & a_2 & a_3 & a_4 \\ b_1 & b_2 & b_3 & b_4 \\ c_1 & c_2 & c_3 & c_4 \\ d_1 & d_2 & d_3 & d_4 \end{vmatrix}$$

will be a sum of 24 terms of the form $\pm a_i b_j c_k d_l$ where $ijkl$ runs over the 24 arrangements of the digits 1234, namely, 1234, 1243, 1324, 1342, 1423, 1432, 2134, 2143, ..., 4321. The problem is to determine the signs of these 24 terms in such a way that $|A| = 0$ whenever two adjacent rows or two adjacent columns of A are equal. If one notices that the formula in the 3×3 case can be written

$$\begin{vmatrix} a_1 & a_2 & a_3 \\ b_1 & b_2 & b_3 \\ c_1 & c_2 & c_3 \end{vmatrix} = a_1 \begin{vmatrix} b_2 & b_3 \\ c_2 & c_3 \end{vmatrix} - b_1 \begin{vmatrix} a_2 & a_3 \\ c_2 & c_3 \end{vmatrix} + c_1 \begin{vmatrix} a_2 & a_3 \\ b_2 & b_3 \end{vmatrix} \tag{1}$$

and that the 2×2 case follows a similar pattern, one is led to the guess

$$\begin{vmatrix} a_1 & a_2 & a_3 & a_4 \\ b_1 & b_2 & b_3 & b_4 \\ c_1 & c_2 & c_3 & c_4 \\ d_1 & d_2 & d_3 & d_4 \end{vmatrix} = a_1 \begin{vmatrix} b_2 & b_3 & b_4 \\ c_2 & c_3 & c_4 \\ d_2 & d_3 & d_4 \end{vmatrix} - b_1 \begin{vmatrix} a_2 & a_3 & a_4 \\ c_2 & c_3 & c_4 \\ d_2 & d_3 & d_4 \end{vmatrix} + c_1 \begin{vmatrix} a_2 & a_3 & a_4 \\ b_2 & b_3 & b_4 \\ d_2 & d_3 & d_4 \end{vmatrix} - d_1 \begin{vmatrix} a_2 & a_3 & a_4 \\ b_2 & b_3 & b_4 \\ c_2 & c_3 & c_4 \end{vmatrix} \tag{2}$$

for a formula that will work in the 4×4 case.

That this formula does work can be seen as follows. As in the 3×3 case, what is to be shown is that if two adjacent rows or two adjacent columns of a 4×4 matrix A are equal, then the value for $|A|$ given by formula (2) is zero. If two adjacent columns of A other than the first two are equal, then each of the four 3×3 determinants in formula (2) contains equal adjacent columns and therefore all four determinants are zero because it has already been shown that 3×3 determinants have this property. Therefore, the value for $|A|$ given by formula (2) is zero in this case. If any two adjacent rows of A are equal, then *two* of the 3×3 determinants in the formula are zero and the remaining two are equal but have opposite signs, so again the value of $|A|$ given by formula (2) is zero. For example, if $c_1 = d_1$, $c_2 = d_2$, $c_3 = d_3$, $c_4 = d_4$, the number given by formula (2) is

$$a_1 \begin{vmatrix} b_2 & b_3 & b_4 \\ c_2 & c_3 & c_4 \\ c_2 & c_3 & c_4 \end{vmatrix} - b_1 \begin{vmatrix} a_2 & a_3 & a_4 \\ c_2 & c_3 & c_4 \\ c_2 & c_3 & c_4 \end{vmatrix} + c_1 \begin{vmatrix} a_2 & a_3 & a_4 \\ b_2 & b_3 & b_4 \\ c_2 & c_3 & c_4 \end{vmatrix} - c_1 \begin{vmatrix} a_2 & a_3 & a_4 \\ b_2 & b_3 & b_4 \\ c_2 & c_3 & c_4 \end{vmatrix} =$$

$$a_1 \cdot 0 - b_1 \cdot 0 + c_1 \begin{vmatrix} a_2 & a_3 & a_4 \\ b_2 & b_3 & b_4 \\ c_2 & c_3 & c_4 \end{vmatrix} - c_1 \begin{vmatrix} a_2 & a_3 & a_4 \\ b_2 & b_3 & b_4 \\ c_2 & c_3 & c_4 \end{vmatrix} = 0.$$

In the remaining case, that in which $a_1 = a_2$, $b_1 = b_2$, $c_1 = c_2$, $d_1 = d_2$, the proof that (2) again gives $|A|$ the value zero is slightly more complicated. If formula (1) is used to expand each of the 3×3 determinants as a sum of 3 terms, the result is a sum of 12 terms

$$\begin{vmatrix} a_1 & a_2 & a_3 & a_4 \\ b_1 & b_2 & b_3 & b_4 \\ c_1 & c_2 & c_3 & c_4 \\ d_1 & d_2 & d_3 & d_4 \end{vmatrix} =$$

$$a_1 b_2 \begin{vmatrix} c_3 & c_4 \\ d_3 & d_4 \end{vmatrix} - a_1 c_2 \begin{vmatrix} b_3 & b_4 \\ d_3 & d_4 \end{vmatrix} + a_1 d_2 \begin{vmatrix} b_3 & b_4 \\ c_3 & c_4 \end{vmatrix} - \cdots - d_1 c_2 \begin{vmatrix} a_3 & a_4 \\ b_3 & b_4 \end{vmatrix}.$$

That these 12 terms sum to zero when $a_1 = a_2, b_1 = b_2, c_1 = c_2, d_1 = d_2$ follows from the fact that they then consist of just 6 different terms, each of which occurs once with the sign $+$ and once with the sign $-$. For example, exactly 2 of the 12 terms are $\pm b_1 d_1 \begin{vmatrix} a_3 & a_4 \\ c_3 & c_4 \end{vmatrix}$, one coming from

$$-b_1 \begin{vmatrix} a_1 & a_3 & a_4 \\ c_1 & c_3 & c_4 \\ d_1 & d_3 & d_4 \end{vmatrix} \text{ and the other from } -d_1 \begin{vmatrix} a_1 & a_3 & a_4 \\ b_1 & b_3 & b_4 \\ c_1 & c_3 & c_4 \end{vmatrix}.$$

The first of these has the sign $-$ in the expansion of $|A|$, and the second has the sign $+$. That the signs are opposite as well for the five other pairs of terms in the expansion is easily checked. (See the proof of the general case below.)

Therefore, for any 4×4 matrix A and any 4×4 tilt T, the number defined by formula (2) has the two properties $|TA| = |A|$, $|AT| = |A|$, which implies, in the same way as in the previous cases:

Theorem. *The only determinant of a* 4×4 *matrix A is the number* $|A|$ *given by formula (2) above.*

5 *The General Case*

The above arguments in the cases $n = 2, 3, 4$ are reviewed and recapitulated in the following proof of the general case:

Theorem. *The only determinant of an $n \times n$ matrix A is the number given by the inductive formula which generalizes formulas (1) and (2) of the last section.*

Specifically, the formula for the determinant $|A|$ of an $n \times n$ matrix is a sum of n terms, in which the ith term is $(-1)^{i-1}$ times the ith entry of the first column of A times the determinant of the $(n-1) \times (n-1)$ matrix obtained by deleting the first column and the ith row of A.

Since each $(n-1) \times (n-1)$ determinant in this formula is defined—by the same definition—as a sum of $n-1$ terms containing $(n-2)\times(n-2)$ determinants, each of which is defined in turn as a sum of $n-2$ terms, and so forth, the formula for $|A|$ consists, ultimately, of a sum of $n(n-1)(n-2)\cdots 3 \cdot 2 \cdot 1 = n!$ terms, each of which is plus or minus a product of n entries of A in which there is one entry from each row as well as one entry from each column.

Proof. Assume the theorem is known to be true for $(n-1) \times (n-1)$ matrices. If two adjacent columns of an $(n-1) \times (n-1)$ matrix are equal, the determinant of the matrix is zero, because such a matrix is a left zero divisor (multiplication of it on the right by an $(n-1) \times 1$ matrix whose entries are all zero except for consecutive entries $1, -1$ in the rows corresponding to the equal columns gives zero), so by Proposition 2 of Section 5 of the last chapter its determinant is zero. Therefore, if two adjacent columns other than the first two columns of an $n \times n$ matrix A are equal, the value of $|A|$ given by the formula is zero because the $(n-1) \times (n-1)$ determinants in all n terms are zero. Similarly, if two adjacent rows of an $(n-1) \times (n-1)$ matrix are equal, the determinant of the matrix is zero. Therefore, if two adjacent rows of an $n \times n$ matrix A are equal, all but two of the n terms in the formula for $|A|$ are equal to zero, and the two remaining terms are consecutive terms with opposite signs and the same absolute value. Therefore the formula gives $|A| = 0$ in this case as well, so it has been shown that the formula gives $|A| = 0$ whenever A has two adjacent rows equal or two adjacent columns other than the first two equal. Finally, that $|A| = 0$ also when the first two columns of A are equal can be proved as follows.

When the formula for an $(n-1) \times (n-1)$ determinant as a sum of $n-1$ terms each containing an $(n-2) \times (n-2)$ determinant is used in the formula for $|A|$ (where A is an $n \times n$ matrix), the result is an expansion of $|A|$ as a sum of $n(n-1)$ terms, each of which is plus or minus a product of an entry of the first column of A, an entry of the second column of A from a different row, and the determinant of the $(n-2) \times (n-2)$ matrix obtained by deleting the first two columns of A and deleting the two rows from which the entries in front were selected. It is clear that each of the $n(n-1)/2$ ways of choosing two rows of A gives rise to exactly two such terms; what is to be shown is that the signs of these two terms are opposite. Say the two selected rows are the ith and the jth, where $i < j$. The two terms are $\pm m_i m_j D_{ij}$, where m_i and m_j are the ith and jth entries, respectively, of the first column of A (which is assumed equal to the second column of A) and where D_{ij} is the determinant of the $(n-2) \times (n-2)$ matrix obtained by deleting rows i and j and columns 1 and 2 of A. One of these two terms comes from the term $(-1)^{i-1} m_i D_i$ of the formula for $|A|$, where D_i is the determinant of the $(n-1) \times (n-1)$ matrix obtained by deleting the ith row and first column of A, and the other comes from the term $(-1)^{j-1} m_j D_j$. When D_i is written as a sum of $n-1$ terms using the formula, the term containing m_j has the sign $(-1)^{j-2}$ because, by virtue of the assumption that $i < j$ and the fact that the ith row is omitted from D_i, it is the $(j-1)$st term of the sum; on the other hand, when D_j is expanded in this way, the term containing m_i has the sign $(-1)^{i-1}$ because it is the ith term of this sum. Therefore, the two terms are $(-1)^{i-1} m_i (-1)^{j-2} m_j D_{ij}$ and $(-1)^{j-1} m_j (-1)^{i-1} m_i D_{ij}$, which shows that their sum is zero. Therefore, $|A| = 0$ for any $n \times n$ matrix in which two adjacent rows are equal or two adjacent columns are equal.

Multiplication of a row or column of A by -1 multiplies $|A|$ by -1 because $|A|$ is a sum of $n!$ terms in which each term is a product containing exactly one factor from the affected row or column, so each of the $n!$ terms is multiplied by -1. Therefore, $|A| = 0$ also holds for any $n \times n$ matrix A in which some column is the negative of an adjacent column or some row is the negative of an adjacent row.

As in formula (1) of Section 3, if A is any $n \times n$ matrix and T is any $n \times n$ tilt, then $|AT| = |A| + |B|$ where B is a matrix in which one column is plus or minus an adjacent column. ($|AT|$ is a sum of $n!$ terms, each of which is a product of $n-1$ entries of A and a sum or difference of two entries of A. Therefore, $|AT|$ is a sum of $2n!$ terms, each of which is plus or minus a product of n entries of A; $n!$ of these are the terms of $|A|$ while the remaining $n!$ terms are $|B|$ for an $n \times n$ matrix B of the type described.) Since, as was just shown, $|B| = 0$ for such a matrix B, it follows that $|AT| = |A|$. In a similar way, $|TA| = |A|$.

If A is equivalent to a diagonal matrix A'', then, on the one hand, $|A''|$ is equal to the product of the diagonal entries of A'', while, on the other hand, each matrix A' that occurs in the chain of linked matrices joining A to A'' satisfies $|A| = |A'|$. Therefore, the product of the diagonal entries of A'' is $|A|$, as was to be shown.

6 *The Determinant of a Product*

As has now been shown, it is valid to call a determinant of a square matrix A *the* determinant of A. The determinant of A will be denoted $|A|$.

Theorem. *The determinant of the product of two $n \times n$ matrices is equal to the product of their determinants. In symbols,* $|AB| = |A||B|$.

Proof. Consider first the case in which A is diagonal, say with diagonal entries $a_1, a_2, \ldots, a_n$ in that order. Then AB is obtained from B by multiplying the first row by a_1, the second row by $a_2, \ldots$, and the last row by a_n. If $|AB|$ and $|B|$ are expanded as sums of $n!$ terms, each of which is plus or minus a product of n entries of the matrix in question, then, because each term contains exactly one entry from each row, $|AB|$ is obtained from $|B|$ by multiplying each term by $a_1 a_2 \cdots a_n$. Since $|A| = a_1 a_2 \cdots a_n$, it follows that $|AB| = |A||B|$ in this case.

An arbitrary $n \times n$ matrix A is equivalent to a diagonal matrix D. Therefore, A can be written in the form $A = MDN$ where M and N are unimodular and D is diagonal. Then $AB = MDNB$. Since equivalent matrices have the same determinant, it follows that $|AB| = |DNB|$. By what was just shown, $|DNB| = |D||NB|$. Since $|D| = |A|$ (because D is equivalent to A) and $|NB| = |B|$ (because NB is equivalent to B), the desired conclusion $|AB| = |A||B|$ follows.

7 The Evaluation of Determinants

The evaluation of the determinant of a large square matrix is, as a general rule, a long computation. Because $n!$ is a very large number even when n is just 7 or 8, direct use of the expression of the determinant of an $n \times n$ matrix as a sum of $n!$ terms is out of the question for large n. A far more effective way to evaluate determinants is to use row or column additions or subtractions to find an equivalent matrix in which one row or column contains only one nonzero entry and to use the following proposition to reduce the problem to the evaluation of a smaller determinant.

Proposition. *If A is a square matrix in which the entry a_{ij} in the ith row of the jth column is the only nonzero entry in the ith row, then* $|A| = (-1)^{i+j} a_{ij} |A_{ij}|$, *where A_{ij} is the $(n-1) \times (n-1)$ matrix obtained by deleting the ith row and jth column of A. The same formula holds if a_{ij} is the only nonzero entry in the jth column.*

Proof. If $j = 1$, the Proposition follows immediately from the formula for $|A|$. It will suffice, therefore, to show that the truth of the Proposition for one value of j implies its truth for $j + 1$. Let P_j be the matrix obtained by interchanging columns j and $j + 1$ of I_n (where $j < n$). The equivalence of 2×2 matrices given by

$$\begin{bmatrix} 0 & 1 \\ 1 & 0 \end{bmatrix} \sim \begin{bmatrix} 1 & 1 \\ 1 & 0 \end{bmatrix} \sim \begin{bmatrix} 1 & 0 \\ 1 & -1 \end{bmatrix} \sim \begin{bmatrix} 1 & 0 \\ 0 & -1 \end{bmatrix}$$

can be applied to rows j and $j+1$ of columns j and $j+1$ of P_j to conclude that P_j is equivalent to the diagonal matrix whose $(j+1)$st diagonal entry is -1 and whose remaining diagonal entries are 1. Since the determinant of the latter matrix is -1, this proves that $|P_j| = -1$. Therefore, $|AP_j| = -|A|$ by the theorem of the last section. But, as is easily seen, AP_j is the matrix obtained by interchanging columns j and $j+1$ of A. When $a_{i,j+1}$ is the only nonzero entry in its row (or column), then it is the entry in the ith row of the jth column of AP_j and is the only nonzero entry in its row (or, respectively, its column). By the inductive hypothesis that the Proposition is true for j, it follows that $|AP_j| = (-1)^{i+j} a_{i,j+1}|A_{i,j+1}|$, because the matrix $A_{i,j+1}$ obtained by deleting row i and column $j+1$ of A is the same as the matrix obtained by deleting row i and column j of AP_j. Therefore, $-|A| = (-1)^{i+j} a_{i,j+1}|A_{i,j+1}|$, which is to say $|A| = (-1)^{i+j+1} a_{i,j+1}|A_{i,j+1}|$, as was to be shown.

For example, the determinant of the matrix

$$\begin{bmatrix} 9 & 3 & 5 \\ 11 & 5 & 5 \\ 16 & 6 & 8 \end{bmatrix}$$

can be found by subtracting the first row from the second and then subtracting the second column from the first to find

$$\begin{vmatrix} 9 & 3 & 5 \\ 11 & 5 & 5 \\ 16 & 6 & 8 \end{vmatrix} = \begin{vmatrix} 9 & 3 & 5 \\ 2 & 2 & 0 \\ 16 & 6 & 8 \end{vmatrix} = \begin{vmatrix} 6 & 3 & 5 \\ 0 & 2 & 0 \\ 10 & 6 & 8 \end{vmatrix} = (-1)^{2+2} 2 \begin{vmatrix} 6 & 5 \\ 10 & 8 \end{vmatrix}.$$

Subtraction of the second column from the first, followed by two subtractions of the first row from the second in the 2×2 determinant in this equation then gives

$$= 2\begin{vmatrix} 1 & 5 \\ 2 & 8 \end{vmatrix} = 2\begin{vmatrix} 1 & 5 \\ 0 & -2 \end{vmatrix} = 2(-1)^{1+1} 1 \cdot (-2) = -4$$

as the value of this 3×3 determinant.

Similarly, the evaluation of an $n \times n$ determinant can always be reduced to the evaluation of an $(n-1) \times (n-1)$ determinant, thence to the evaluation of an $(n-2) \times (n-2)$ determinant, and so forth, until only a 1×1 determinant is needed.

In evaluating determinants by this method, it is useful to know that adding a row to *any* other row—not just an adjacent row—gives an equivalent matrix and therefore leaves the determinant unchanged. The same is true of row subtractions, column additions, and column subtractions. (See Exercise 7.)

In the evaluation of determinants it is sometimes useful also to make use of the fact that (by the formula for $|A|$ as a sum of $n!$ terms) a common factor of the entries of any row or column can be taken out in front of the determinant. For example, in the above 3×3 determinant, the factor 2 in the last row can be taken out in front to find

$$\begin{vmatrix} 9 & 3 & 5 \\ 11 & 5 & 5 \\ 16 & 6 & 8 \end{vmatrix} = 2 \begin{vmatrix} 9 & 3 & 5 \\ 11 & 5 & 5 \\ 8 & 3 & 4 \end{vmatrix}.$$

Further simplifications give

$$2 \begin{vmatrix} 9 & 3 & 5 \\ 11 & 5 & 5 \\ 8 & 3 & 4 \end{vmatrix} = 2 \begin{vmatrix} 1 & 0 & 1 \\ 11 & 5 & 5 \\ 8 & 3 & 4 \end{vmatrix} = 2 \begin{vmatrix} 0 & 0 & 1 \\ 6 & 5 & 5 \\ 4 & 3 & 4 \end{vmatrix} = 2(-1)^{1+3} \cdot 1 \begin{vmatrix} 6 & 5 \\ 4 & 3 \end{vmatrix} =$$

$$= 2 \begin{vmatrix} 2 & 2 \\ 4 & 3 \end{vmatrix} = 2 \begin{vmatrix} 2 & 0 \\ 4 & -1 \end{vmatrix} = 2 \cdot (-1)^{1+1} \cdot 2 \cdot (-1) = -4.$$

(At the first step, the last row is subtracted from the first. At the second, the last column is subtracted from the first. Next, use is made of the fact that the top entry of the last column is the only nonzero entry in its row. Then the second row is subtracted from the first. Then the first column is subtracted from the second. Finally, use is made of the fact that the top entry in the first column is the only nonzero entry in its row.)

Examples

The reduction of $A = \begin{bmatrix} 1 & 2 & 3 \\ 4 & 5 & 6 \\ 7 & 8 & 0 \end{bmatrix}$ to diagonal form gives

$$\begin{bmatrix} 1 & 2 & 3 \\ 4 & 5 & 6 \\ 7 & 8 & 0 \end{bmatrix} \sim \begin{bmatrix} 1 & 2 & 1 \\ 4 & 5 & 1 \\ 7 & 8 & -8 \end{bmatrix} \sim \begin{bmatrix} 1 & 1 & 1 \\ 4 & 4 & 1 \\ 7 & 16 & -8 \end{bmatrix} \sim \begin{bmatrix} 1 & 1 & 0 \\ 4 & 4 & -3 \\ 7 & 16 & -24 \end{bmatrix}$$

$$\sim \begin{bmatrix} 1 & 0 & 0 \\ 4 & 0 & -3 \\ 7 & 9 & -24 \end{bmatrix} \sim \begin{bmatrix} 1 & 0 & 0 \\ 4 & 0 & -3 \\ 3 & 9 & -21 \end{bmatrix} \sim \begin{bmatrix} 1 & 0 & 0 \\ 1 & -9 & 18 \\ 3 & 9 & -21 \end{bmatrix} \sim \begin{bmatrix} 1 & 0 & 0 \\ 1 & -9 & 18 \\ 0 & 36 & -75 \end{bmatrix}$$

$$\sim \begin{bmatrix} 1 & 0 & 0 \\ 0 & -9 & 18 \\ 0 & 36 & -75 \end{bmatrix} \sim \begin{bmatrix} 1 & 0 & 0 \\ 0 & 9 & 18 \\ 0 & -39 & -75 \end{bmatrix} \sim \begin{bmatrix} 1 & 0 & 0 \\ 0 & 9 & 0 \\ 0 & -39 & 3 \end{bmatrix} \sim \begin{bmatrix} 1 & 0 & 0 \\ 0 & 9 & 0 \\ 0 & 6 & 3 \end{bmatrix}$$

$$\sim \begin{bmatrix} 1 & 0 & 0 \\ 0 & 3 & -3 \\ 0 & 6 & 3 \end{bmatrix} \sim \begin{bmatrix} 1 & 0 & 0 \\ 0 & 3 & 0 \\ 0 & 6 & 9 \end{bmatrix} \sim \begin{bmatrix} 1 & 0 & 0 \\ 0 & 3 & 0 \\ 0 & 0 & 9 \end{bmatrix}$$

(where repetitions of the same operation have been condensed into one step). Therefore, $1 \cdot 3 \cdot 9$ is the determinant of A. This determinant can be found without any reductions by using the formula for 3×3 determinants,

$$\begin{vmatrix} 1 & 2 & 3 \\ 4 & 5 & 6 \\ 7 & 8 & 0 \end{vmatrix} = 1 \cdot \begin{vmatrix} 5 & 6 \\ 8 & 0 \end{vmatrix} - 4 \cdot \begin{vmatrix} 2 & 3 \\ 8 & 0 \end{vmatrix} + 7 \cdot \begin{vmatrix} 2 & 3 \\ 5 & 6 \end{vmatrix} = (0-48) - 4(0-24) + 7(12-15) = 27.$$

The purpose of the formula is the *theoretical* one of showing that the answer 27 does not depend on the choice of a reduction of A. It is rarely a good way to evaluate the determinant of a matrix. For example, the simple reductions on the first line above give

$$\begin{vmatrix} 1 & 2 & 3 \\ 4 & 5 & 6 \\ 7 & 8 & 0 \end{vmatrix} = \begin{vmatrix} 1 & 0 & 0 \\ 4 & 0 & -3 \\ 7 & 9 & -24 \end{vmatrix} = \begin{vmatrix} 0 & -3 \\ 9 & -24 \end{vmatrix} = 27.$$

Even easier, subtraction of the first row twice from the second, followed by reductions of the resulting 2×2 matrix gives

$$\begin{vmatrix} 1 & 2 & 3 \\ 4 & 5 & 6 \\ 7 & 8 & 0 \end{vmatrix} = \begin{vmatrix} 1 & 2 & 3 \\ 2 & 1 & 0 \\ 7 & 8 & 0 \end{vmatrix} = (-1)^{1+3}3\begin{vmatrix} 2 & 1 \\ 7 & 8 \end{vmatrix} = 3\begin{vmatrix} 0 & 1 \\ -9 & 8 \end{vmatrix} = 3(-1)^{1+2}1 \cdot (-9) = 27.$$

Let $B = \begin{bmatrix} 2 & 0 & 1 \\ -1 & -1 & -3 \\ 1 & 2 & 2 \end{bmatrix}$. The following calculations verify the formula $|AB| = |A||B|$ of Section 6.

$$|B| = \begin{vmatrix} 2 & 0 & 1 \\ -1 & -1 & -3 \\ 1 & 2 & 2 \end{vmatrix} = \begin{vmatrix} 2 & 0 & 1 \\ -1 & -1 & -3 \\ -1 & 0 & -4 \end{vmatrix}$$

$$= (-1)^4(-1)\begin{vmatrix} 2 & 1 \\ -1 & -4 \end{vmatrix} = -\begin{vmatrix} 0 & -7 \\ -1 & -4 \end{vmatrix} = 7.$$

$$AB = \begin{bmatrix} 1 & 2 & 3 \\ 4 & 5 & 6 \\ 7 & 8 & 0 \end{bmatrix}\begin{bmatrix} 2 & 0 & 1 \\ -1 & -1 & -3 \\ 1 & 2 & 2 \end{bmatrix} = \begin{bmatrix} 3 & 4 & 1 \\ 9 & 7 & 1 \\ 6 & -8 & -17 \end{bmatrix}$$

$$|AB| = 3\begin{vmatrix} 1 & 4 & 1 \\ 3 & 7 & 1 \\ 2 & -8 & -17 \end{vmatrix} = 3\begin{vmatrix} -2 & -3 & 0 \\ 3 & 7 & 1 \\ 53 & 111 & 0 \end{vmatrix}$$

$$= 3(-1)^5\begin{vmatrix} -2 & -3 \\ 53 & 111 \end{vmatrix} = -3\begin{vmatrix} -2 & -1 \\ 53 & 58 \end{vmatrix}$$

$$= -3\begin{vmatrix} -1 & -1 \\ -5 & 58 \end{vmatrix} = -3\begin{vmatrix} -1 & 0 \\ -5 & 63 \end{vmatrix} = 3 \cdot 63 = 27 \cdot 7 = |A||B|.$$

Exercises

1 Evaluate the determinant

$$\begin{vmatrix} 3 & -3 & 6 & 0 & 0 \\ 3 & 0 & 9 & 6 & 2 \\ 3 & 1 & 11 & 12 & 3 \\ 1 & 3 & 4 & -3 & 3 \\ 1 & -1 & 0 & -5 & -2 \end{vmatrix}.$$

(Feel free to use the fact that rows or columns can be added to or subtracted from others—even non-adjacent ones—without changing the determinant. See Exercise 7.)

2 Evaluate the determinant

$$\begin{vmatrix} 0 & 1 & 4 & 4 & -4 \\ 0 & 1 & 1 & 1 & 4 \\ 2 & 1 & -1 & 0 & 2 \\ 1 & 2 & 3 & 4 & -3 \\ 1 & -1 & 1 & 1 & 5 \end{vmatrix}.$$

3 Let

$$A = \begin{bmatrix} 2 & 0 & 1 & 0 \\ 0 & 3 & 0 & 2 \\ 2 & 0 & 0 & 3 \\ 1 & 2 & 0 & 0 \end{bmatrix} \quad \text{and} \quad B = \begin{bmatrix} 0 & 3 & 2 & 0 \\ 2 & 0 & 0 & 1 \\ 1 & 0 & 5 & 0 \\ 0 & 2 & 0 & 3 \end{bmatrix}.$$

Find AB, $|A|$, $|B|$, and $|AB|$.

4 A **permutation matrix** is a square matrix in which each row contains one 1 and the rest zeros, and each column contains one 1 and the rest zeros. For example, there are just two 2×2 permutation matrices $\begin{bmatrix} 1 & 0 \\ 0 & 1 \end{bmatrix}$ and $\begin{bmatrix} 0 & 1 \\ 1 & 0 \end{bmatrix}$.

(a) Find the six 3×3 permutation matrices, and find the determinant of each.

(b) How many $n \times n$ permutation matrices are there? Show that, when $n > 1$, exactly half of them have determinant 1. What are the determinants of the other half?

5 Use the notion of permutation matrices and their determinants to describe a formula for the determinant of an $n \times n$ matrix.

6 Show that interchange of two adjacent rows (or two adjacent columns) in a square matrix changes the sign of its determinant. Show that interchange of any two rows (or columns) in a square matrix changes the sign of its determinant. Show without computation that

$$\begin{vmatrix} 2 & 3 & 5 \\ 2 & 7 & -1 \\ 2 & 3 & 5 \end{vmatrix} = 0.$$

7 Show that adding any row (or column) of a square matrix to any other, or subtracting it from any other, leaves the determinant unchanged.

8 The **transpose** of an $m \times n$ matrix is the $n \times m$ matrix obtained by writing the rows as columns (see Chapter 7). Show that the determinant of any square matrix is the same as the determinant of its transpose.

9 Prove that a matrix with determinant 1 is unimodular.

10 Let A be the matrix obtained by replacing the jth column of I_n with the column whose entries are $c_1, c_2, \ldots, c_n$. Find $|A|$.

11 If integers x, y, z satisfy

$$\begin{bmatrix} 2 & 1 & 4 \\ 3 & -2 & 5 \\ 1 & 2 & 3 \end{bmatrix} \begin{bmatrix} x \\ y \\ z \end{bmatrix} = \begin{bmatrix} 3 \\ 0 \\ 4 \end{bmatrix}.$$

then they also satisfy

$$\begin{bmatrix} 2 & 1 & 4 \\ 3 & -2 & 5 \\ 1 & 2 & 3 \end{bmatrix} \begin{bmatrix} 1 & x & 0 \\ 0 & y & 0 \\ 0 & z & 1 \end{bmatrix} = \begin{bmatrix} 2 & 3 & 4 \\ 3 & 0 & 5 \\ 1 & 4 & 3 \end{bmatrix}.$$

If one takes the determinant of this matrix and makes use of the fact that the determinant of a product is the product of the determinants and of the fact that the second factor on the left has determinant y, one finds a formula for y as a quotient of two 3×3 determinants. Use these observations to find y; then find x and z by a similar method.

12 *Cramer's Rule.* Using the method of the previous exercise, show that each component x_1, x_2, $\ldots$, x_n of the solution of an $n \times n$ system of equations

$$\begin{aligned} a_{11}x_1 + a_{12}x_2 + \cdots + a_{1n}x_n &= c_1 \\ a_{21}x_1 + a_{22}x_2 + \cdots + a_{2n}x_n &= c_2 \\ \vdots \quad\quad \vdots \quad\quad\quad\quad \vdots \quad &\quad \vdots \\ a_{n1}x_1 + a_{n2}x_2 + \cdots + a_{nn}x_n &= c_n \end{aligned}$$

can be expressed as a quotient of two determinants, provided the determinant of the $n \times n$ matrix of coefficients a_{ij} on the right is not zero. Describe explicitly the quotient which gives the value of x_i for each i. This formula for the solution of linear equations is known as Cramer's Rule. It has important theoretical uses, but its computational usefulness is nil because the evaluation of the determinants is so long.

13 Use Cramer's rule to give a formula for the entry in the ith row of the jth column of the inverse of an invertible $n \times n$ matrix.

14 Let $-I_n$ denote the matrix obtained by reversing the signs of all entries in I_n. When is I_n equivalent to $-I_n$?

15 Use the ideas of the Supplementary Unit of Chapter 5 to prove, without using the theory of determinants, that *except for the sign* a square matrix has only one determinant, that is, a square matrix has at most two determinants, which differ in sign only.

Supplementary Unit: Exterior Algebra

Another approach to the theory of determinants is through the use of so-called exterior algebra.

The discussion of exterior algebra given below assumes that the reader is familiar with the algebra of polynomials in any number of variables (more properly called indeterminates) with integer coefficients. For purposes of illustration, we will consider polynomials in three variables x, y, z with integer coefficients, but the same rules apply to the case of any number of variables represented by any set of symbols. Such polynomials are written in the familiar way as sums of monomials, where a monomial is an integer times a product of powers of the variables. They are added and multiplied in the familiar way, making use of the associative, commutative, and distributive laws of addition and multiplication. The symbol 0 represents, in this context, the "polynomial" which is a sum of *no* monomials.

The algebra of polynomials in x, y, z is to be regarded here as a formal system of calculation, in which it is understood how to present polynomials, how to add them, how to multiply them, and, most importantly, *how to determine whether two presentations of polynomials are to be regarded as equal.* Thus, laws like $x + y = y + x$ and $xy = yx$ are to be regarded as expressing nothing more than the rules governing calculation with polynomials—$x + y$ and $y + x$ are not identical, but in the algebra of polynomials it is agreed to regard them as different presentations of the same polynomial.

Exterior algebra is based on a modification of the algebra of polynomials in which *products of variables are no longer assumed to commute.* To avoid confusion with the ordinary algebra of polynomials, a wedge $\wedge$ will be written between two variables to represent their product. Thus, in this new algebra, which we will call the algebra of noncommutative polynomials (although addition is still commutative), monomials are expressions of the form $3x \wedge y \wedge x$ or $2y \wedge x \wedge x$, that is, an integer (possibly 1) times some product of variables (possibly a product of none) in which wedges are written between successive variables. Polynomials are sums of such monomials. Two polynomials are added by the obvious method of combining the two sums. Multiplication of polynomials requires more comment. First, two monomials are multiplied by multiplying the integers in front of them (so that integers commute with variables) and following that integer by the product of the variables in each monomial, in the same order, separated by wedges—for example, $(2x \wedge z) \wedge (3z \wedge y \wedge x) = 6x \wedge z \wedge z \wedge y \wedge x$. Because multiplication of monomials need not be commutative, it is indicated here by a wedge rather than simple juxtaposition. The product of two polynomials is then found by using the distributive law to write it as a sum of products of monomials—for example $(3 + x) \wedge (y + x \wedge z) = 3y + 3x \wedge z + x \wedge y + x \wedge x \wedge z$.

Once again, this algebra of noncommutative polynomials consists in knowing how to present such a polynomial, how to add two of them, how to multiply two of them (use $\wedge$ to indicate multiplication of polynomials as a reminder that the operation need not be commutative), and how to show that two are equal. Briefly put, two are equal if one can be transformed into the other using the usual rules of associativity, commutativity of addition, and distributivity, but without using commutativity of the multiplication of variables.

In the following exercises, we will take for granted the following statement about the algebra of noncommutative polynomials: Every noncommutative polynomial can be written as a sum of monomials in which no two monomials contain the same two products of variables. A noncommutative polynomial in this form is equal to 0 only if it is identical to 0, that is, only if it is a sum of *no* terms.

The algebra of noncommutative polynomials becomes **exterior algebra** when the further computational rule is added that *variables anticommute,* that is, $x \wedge y = -y \wedge x$, $x\wedge z = -z\wedge x$, $y\wedge z = -z\wedge y$, $x\wedge x = 0$, $y\wedge y = 0$, $z\wedge z = 0$. (The last three identities are natural concomitants of anticommutativity of variables, because $x \wedge x = -x \wedge x$ implies $2x \wedge x = 0$, but formally they must be listed as separate rules of the algebra because nothing in the rules given so far permits one to conclude from $2x \wedge x = 0$ that $x \wedge x = 0$.)

An **exterior polynomial** is represented in the same way as a polynomial in noncommutative algebra. Two exterior polynomials are to be regarded as **equal** if one can be transformed into the other using the rules of noncommutative algebra and the rules $x \wedge y = -y \wedge x$, $x \wedge x = 0$, etc. given above.

1 Let an order be established for the variables in the exterior algebra under consideration. (In the main example, the variables will be taken to be x, y, z, in that order.) Define an **atom** be an exterior polynomial which is a product of *distinct* variables in which the variables appear *in order.* By convention, 1 is considered to be an atom because it is a product of *no* variables. (Thus, in the main example, the atoms are 1, x, y, z, $x \wedge y$, $x \wedge z$, $y \wedge z$, and $x \wedge y \wedge z$.) (a) When there are n variables, how many atoms are there? The **lexicographic order** of the atoms is established by the rule that one atom **precedes** another if (1) it contains fewer variables, or (2) it contain the same number of variables as the other atom, and the first variable that is not in both atoms is in it. (b) In the case of four variables a, b, c, d, list the atoms in lexicographic order. An exterior polynomial is in **canonical form** if it is a sum of integer multiples of atoms, and if the atoms occur in the sum in lexicographic order. By convention, the polynomial 0, which contains no terms, is considered to be in canonical form. (c) Show that every exterior polynomial is equal to one in canonical form.

2 The main difficulty in the theory of exterior algebra lies in giving a method for showing that exterior polynomials are *not* equal. (Similarly, the main difficulty in the theory of equivalence of matrices—solved in Chapter 5 using determinants—lies in giving a method for showing that matrices are *not* equivalent.) The fundamental theorem states that *two exterior polynomials in canonical form are equal only if they are identical.* (a) Show that if two exterior polynomials in canonical form are equal but not identical, then an integer multiple of the product of all the variables in order is equal to zero without the multiplier being zero. (In the three-variable case, then, there is an equation of the form $ax \wedge y \wedge z = 0$ in which a is a nonzero integer.) (b) When there is just one variable, prove that two exterior polynomials in canonical form are equal only if they are identical. (c) Do the same in the case of two variables. (d) Do the same in the case of 3 variables. (e) Do the same in the case of n variables.

3 In exterior algebra, an $m \times n$ matrix A corresponds in a natural way to a $2^m \times 2^n$ matrix $A^\wedge$ in the following way. Suppose, for example, that $A = \begin{bmatrix} 1 & 2 & 3 \\ 4 & 5 & 6 \end{bmatrix}$. Regard A as the matrix of coefficients of a linear substitution, say $u = x + 2y + 3z$, $v = 4x + 5y + 6z$. (See Chapter 1.) The exterior polynomials in two variables u and v have the form $a + bu + cv + du \wedge v$ where a, b, c, and d are integers. The linear substitution expressing u and v in terms of x, y, and z can be used to express any exterior polynomial in u and v in terms of x, y, and z. For example, $u\wedge v = (x+2y+3z)\wedge(4x+5y+6z) = 4x\wedge x+5x\wedge y+6x\wedge z+8y\wedge x+10y\wedge y+12y\wedge z+12z\wedge x+15z\wedge y+18z\wedge z = -3x\wedge y-6x\wedge z-3y\wedge z$. Therefore, the expression of $a+bu+cv+du\wedge v$ in terms of x, y, and z is $a + b(x + 2y + 3z) + c(4x + 5y + 6z) + d(-3x \wedge y - 6x \wedge z - 3y \wedge z)$, or, in matrix form,

$$\begin{bmatrix} a & b & c & d \end{bmatrix} \begin{bmatrix} 1 \\ u \\ v \\ u \wedge v \end{bmatrix} = \begin{bmatrix} a & b & c & d \end{bmatrix} \begin{bmatrix} 1 & 0 & 0 & 0 & 0 & 0 & 0 & 0 \\ 0 & 1 & 2 & 3 & 0 & 0 & 0 & 0 \\ 0 & 4 & 5 & 6 & 0 & 0 & 0 & 0 \\ 0 & 0 & 0 & 0 & -3 & -6 & -3 & 0 \end{bmatrix} \begin{bmatrix} 1 \\ x \\ y \\ z \\ x \wedge y \\ x \wedge z \\ y \wedge z \\ x \wedge y \wedge z \end{bmatrix}.$$

The 4×8 matrix in this equation is $A^\wedge$ for this matrix A.
Find $A^\wedge$ for the following matrices A:

$$\text{(a)}\quad A = \begin{bmatrix} 1 & 0 \\ 1 & 1 \\ 1 & 2 \end{bmatrix} \qquad \text{(b)}\quad A = \begin{bmatrix} 3 & 5 \\ 2 & 4 \end{bmatrix} \qquad \text{(c)}\quad A = \begin{bmatrix} 0 & 3 & 1 \\ 2 & 0 & -1 \\ 2 & 1 & 0 \end{bmatrix}.$$

4 Prove that if A and B are matrices for which AB is defined, then $(AB)^\wedge = A^\wedge B^\wedge$.

5 As the examples show, $A^\wedge$ is made up in a natural way of component matrices, which we will denote $A^{(0)}, A^{(1)}, A^{(2)}, \ldots, A^{(k)}$, where $A^{(0)} = 1$, $A^{(1)} = A$, and where k is the minimum of m and n. (a) How many rows and columns does $A^{(i)}$ have? (b) Show that $(AB)^{(i)} = A^{(i)}B^{(i)}$ whenever i is an integer for which both $A^{(i)}$ and $B^{(i)}$ are defined. (c) Show that if A is an $n \times n$ matrix then $A^{(n)} = |A|$. Conclude that the determinant of a product is the product of the determinants.

Chapter 5

Testing for Equivalence

1 Introduction

THE INTRODUCTION of the notion of equivalence of matrices in Chapter 2, like the introduction of any notion of equivalence in mathematics, raises a natural question: How can one determine whether two matrices are equivalent? An answer is given in this chapter.

2 Strongly Diagonal Matrices

We will say that an $m \times n$ matrix is **strongly diagonal*** if it is diagonal and if, in addition, (1) each diagonal entry after the first is an integer multiple of the preceding diagonal entry, and (2) there are no negative entries, except that in a square matrix the *last* diagonal entry may be negative.

Theorem. *Every matrix is equivalent to a strongly diagonal matrix.*

This theorem will be proved by means of an extension of the algorithm of Chapter 2.

Algorithm: If the matrix is not diagonal, perform the operation determined by rules 1 to 3 and rules $1'$ to $3'$ of Chapter 2, Section 7. If the matrix is diagonal, perform the operation, if any, determined by:

4. If some diagonal entry is not an integer multiple of its predecessor on the diagonal, add row i to row $i - 1$, where the ith diagonal entry is the first diagonal entry that fails to be a multiple of its predecessor.
5. If 4 does not determine an operation and if there is a negative diagonal entry in a column other than the last, subtract column i from column $i + 1$, where the ith diagonal entry is the first negative one.

*Except for the special condition about the sign of the last diagonal entry of a square matrix, a strongly diagonal matrix is what is often called a matrix in *Smith canonical form*.

6. If neither 4 nor 5 determines an operation and if there is a negative diagonal entry in a row other than the last, subtract row i from row $i+1$, where the ith diagonal entry is the first negative one.

If none of 4, 5, or 6 determines an operation, the matrix is strongly diagonal and the algorithm terminates.

This algorithm produces a sequence of linked matrices and terminates with a strongly diagonal matrix equivalent to the original matrix. For example, it produces the sequence

$$\begin{bmatrix} -1 & 0 & 0 & 0 \\ 0 & -2 & 0 & 0 \\ 0 & 0 & 0 & 0 \\ 0 & 0 & 0 & -3 \end{bmatrix} \sim \begin{bmatrix} -1 & 0 & 0 & 0 \\ 0 & -2 & 0 & 0 \\ 0 & 0 & 0 & -3 \\ 0 & 0 & 0 & -3 \end{bmatrix} \sim \begin{bmatrix} -1 & 0 & 0 & 0 \\ 0 & -2 & 0 & 0 \\ 0 & 0 & 3 & -3 \\ 0 & 0 & 3 & -3 \end{bmatrix} \sim$$

$$\sim \begin{bmatrix} -1 & 0 & 0 & 0 \\ 0 & -2 & 0 & 0 \\ 0 & 0 & 3 & 0 \\ 0 & 0 & 3 & 0 \end{bmatrix} \sim \begin{bmatrix} -1 & 0 & 0 & 0 \\ 0 & -2 & 0 & 0 \\ 0 & 0 & 3 & 0 \\ 0 & 0 & 0 & 0 \end{bmatrix} \sim \begin{bmatrix} -1 & 0 & 0 & 0 \\ 0 & -2 & 3 & 0 \\ 0 & 0 & 3 & 0 \\ 0 & 0 & 0 & 0 \end{bmatrix} \sim$$

$$\sim \begin{bmatrix} -1 & 0 & 0 & 0 \\ 0 & 1 & 3 & 0 \\ 0 & 3 & 3 & 0 \\ 0 & 0 & 0 & 0 \end{bmatrix} \sim \begin{bmatrix} -1 & 0 & 0 & 0 \\ 0 & 1 & 2 & 0 \\ 0 & 3 & 0 & 0 \\ 0 & 0 & 0 & 0 \end{bmatrix} \sim \begin{bmatrix} -1 & 0 & 0 & 0 \\ 0 & 1 & 1 & 0 \\ 0 & 3 & -3 & 0 \\ 0 & 0 & 0 & 0 \end{bmatrix} \sim$$

$$\sim \begin{bmatrix} -1 & 0 & 0 & 0 \\ 0 & 1 & 0 & 0 \\ 0 & 3 & -6 & 0 \\ 0 & 0 & 0 & 0 \end{bmatrix} \sim \begin{bmatrix} -1 & 0 & 0 & 0 \\ 0 & 1 & 0 & 0 \\ 0 & 2 & -6 & 0 \\ 0 & 0 & 0 & 0 \end{bmatrix} \sim \begin{bmatrix} -1 & 0 & 0 & 0 \\ 0 & 1 & 0 & 0 \\ 0 & 1 & -6 & 0 \\ 0 & 0 & 0 & 0 \end{bmatrix} \sim$$

$$\sim \begin{bmatrix} -1 & 0 & 0 & 0 \\ 0 & 1 & 0 & 0 \\ 0 & 0 & -6 & 0 \\ 0 & 0 & 0 & 0 \end{bmatrix} \sim \begin{bmatrix} -1 & 1 & 0 & 0 \\ 0 & 1 & 0 & 0 \\ 0 & 0 & -6 & 0 \\ 0 & 0 & 0 & 0 \end{bmatrix} \sim \begin{bmatrix} 0 & 1 & 0 & 0 \\ 1 & 1 & 0 & 0 \\ 0 & 0 & -6 & 0 \\ 0 & 0 & 0 & 0 \end{bmatrix} \sim$$

$$\sim \begin{bmatrix} 1 & 1 & 0 & 0 \\ 2 & 1 & 0 & 0 \\ 0 & 0 & -6 & 0 \\ 0 & 0 & 0 & 0 \end{bmatrix} \sim \begin{bmatrix} 1 & 0 & 0 & 0 \\ 2 & -1 & 0 & 0 \\ 0 & 0 & -6 & 0 \\ 0 & 0 & 0 & 0 \end{bmatrix} \sim \begin{bmatrix} 1 & 0 & 0 & 0 \\ 1 & -1 & 0 & 0 \\ 0 & 0 & -6 & 0 \\ 0 & 0 & 0 & 0 \end{bmatrix} \sim$$

$$\sim \begin{bmatrix} 1 & 0 & 0 & 0 \\ 0 & -1 & 0 & 0 \\ 0 & 0 & -6 & 0 \\ 0 & 0 & 0 & 0 \end{bmatrix} \sim \begin{bmatrix} 1 & 0 & 0 & 0 \\ 0 & -1 & 1 & 0 \\ 0 & 0 & -6 & 0 \\ 0 & 0 & 0 & 0 \end{bmatrix} \sim \begin{bmatrix} 1 & 0 & 0 & 0 \\ 0 & 0 & 1 & 0 \\ 0 & -6 & -6 & 0 \\ 0 & 0 & 0 & 0 \end{bmatrix} \sim$$

$$\sim \begin{bmatrix} 1 & 0 & 0 & 0 \\ 0 & 1 & 1 & 0 \\ 0 & -12 & -6 & 0 \\ 0 & 0 & 0 & 0 \end{bmatrix} \sim \begin{bmatrix} 1 & 0 & 0 & 0 \\ 0 & 1 & 0 & 0 \\ 0 & -12 & 6 & 0 \\ 0 & 0 & 0 & 0 \end{bmatrix} \sim \begin{bmatrix} 1 & 0 & 0 & 0 \\ 0 & 1 & 0 & 0 \\ 0 & -11 & 6 & 0 \\ 0 & 0 & 0 & 0 \end{bmatrix} \sim$$

$$\sim \cdots \sim \begin{bmatrix} 1 & 0 & 0 & 0 \\ 0 & 1 & 0 & 0 \\ 0 & -1 & 6 & 0 \\ 0 & 0 & 0 & 0 \end{bmatrix} \sim \begin{bmatrix} 1 & 0 & 0 & 0 \\ 0 & 1 & 0 & 0 \\ 0 & 0 & 6 & 0 \\ 0 & 0 & 0 & 0 \end{bmatrix}$$

ending with a strongly diagonal matrix.

Proof that the algorithm terminates. Consider a sequence of linked matrices $A_1 \sim A_2 \sim A_3 \sim \cdots$ generated by the algorithm. As was shown in Chapter 2, this sequence must contain a diagonal matrix. Let D_1 be the first diagonal matrix in the sequence. If D_1 is not strongly diagonal, the algorithm calls for an operation of type 4, 5, or 6.

Suppose first that it calls for an operation of type 4, that is, some diagonal entry of the diagonal matrix D_1 is not a multiple of its predecessor on the diagonal. The matrix following D_1 in the sequence results from applying an operation of type 4 to D_1, so it has a nonzero entry in row $i - 1$ of column i for some i and otherwise is diagonal. Subsequent operations change entries only in columns $i - 1$ and i of rows $i - 1$ and i until the next diagonal matrix, call it D_2, is reached. The transition from D_1 to D_2 occurs entirely within this 2×2 matrix and can take just a few forms.

If the $(i - 1)$st and ith diagonal entries of D_1 are 0, c where $c > 0$, the transition is

$$\begin{bmatrix} 0 & 0 \\ 0 & c \end{bmatrix} \sim \begin{bmatrix} 0 & c \\ 0 & c \end{bmatrix} \sim \begin{bmatrix} c & c \\ c & c \end{bmatrix} \sim \begin{bmatrix} c & 0 \\ c & 0 \end{bmatrix} \sim \begin{bmatrix} c & 0 \\ 0 & 0 \end{bmatrix}.$$

If, on the other hand, the diagonal entries in question are 0 and $-c$ where $c > 0$, the transition is

$$\begin{bmatrix} 0 & 0 \\ 0 & -c \end{bmatrix} \sim \begin{bmatrix} 0 & -c \\ 0 & -c \end{bmatrix} \sim \begin{bmatrix} c & -c \\ c & -c \end{bmatrix} \sim \begin{bmatrix} c & 0 \\ c & 0 \end{bmatrix} \sim \begin{bmatrix} c & 0 \\ 0 & 0 \end{bmatrix}.$$

Finally, if the diagonal entries in question are d_1, d_2 where $d_1 \neq 0$ and d_2 is not a multiple of d_1, the transition is

$$\begin{bmatrix} d_1 & 0 \\ 0 & d_2 \end{bmatrix} \sim \begin{bmatrix} d_1 & d_2 \\ 0 & d_2 \end{bmatrix} \sim \cdots \sim \begin{bmatrix} a & 0 \\ b & c \end{bmatrix} \sim \cdots \sim \begin{bmatrix} e & 0 \\ 0 & f \end{bmatrix}$$

where $\begin{bmatrix} a & 0 \\ b & c \end{bmatrix}$ stands for the first matrix in the sequence in which the upper right corner is zero and where e and f are the $(i - 1)$st and the ith diagonal entries in the next diagonal matrix D_2 in the sequence. The equivalence of these matrices implies that $d_1 d_2 = ef$. That e is positive and that $e < |d_1|$ can be seen as follows.

If d_1 and d_2 are both negative, then, since $d_1 \neq d_2$, the matrix following $\begin{bmatrix} d_1 & d_2 \\ 0 & d_2 \end{bmatrix}$ is obtained by subtracting the column containing the lesser of d_1 and d_2 in its first row from the column containing the greater, which creates a positive entry in the first row. Thus, at least one matrix in the sequence prior to $\begin{bmatrix} a & 0 \\ b & c \end{bmatrix}$ has a positive entry in the first row. It follows that all later matrices up to and including $\begin{bmatrix} a & 0 \\ b & c \end{bmatrix}$ have a positive entry in the first row, so $a > 0$. Up to this point, all operations are column operations. Therefore, because a divides both entries in the first row of $\begin{bmatrix} a & 0 \\ b & c \end{bmatrix}$, it also divides both entries in the first row of all these matrices and, in particular, divides d_1 and d_2. Since d_2 is not a multiple of d_1, it follows that $a < |d_1|$. Moreover, since a divides both d_1 and d_2, it divides all entries of all matrices in the above sequence. In particular, $b = qa$ for some integer q. Therefore, the transition from $\begin{bmatrix} a & 0 \\ b & c \end{bmatrix}$ to $\begin{bmatrix} e & 0 \\ 0 & f \end{bmatrix}$ is simply a sequence of additions of the first row to, or subtractions of the first row from, the second row until the matrix becomes diagonal. In short, $e = a$ and $f = c$.

Thus, in the transition from the first diagonal matrix D_1 in the sequence to the second diagonal matrix D_2, *either a zero on the diagonal has moved to a position farther down the diagonal or, for some k, the product of the first k nonzero diagonal entries has decreased in absolute value, while for all other values of k the product of the first k nonzero diagonal entries is unchanged.* Since a diagonal matrix can undergo at most a finite number of alterations of this type, the sequence $A_1 \sim A_2 \sim \cdots$ must contain a diagonal matrix to which rule 4 does not apply, that is, in which each diagonal entry is a multiple of its predecessor on the diagonal.

If the first such matrix is not strongly diagonal, then either rule 5 or rule 6 must apply. Assume first that rule 5 does.

Then $d_i = -c$ where $c > 0$ and $d_{i+1} = qc$ for some integer q. The transition to the next diagonal matrix after the operation of rule 5 is then

$$\begin{bmatrix} -c & c \\ 0 & qc \end{bmatrix} \sim \begin{bmatrix} 0 & c \\ qc & qc \end{bmatrix} \sim \begin{bmatrix} c & c \\ 2qc & qc \end{bmatrix} \sim \begin{bmatrix} c & 0 \\ 2qc & -qc \end{bmatrix}$$

followed by $|2q|$ additions or subtractions using the first row to reduce the first entry of the second row to zero. Thus, this transition changes $\begin{bmatrix} -c & 0 \\ 0 & qc \end{bmatrix}$ to $\begin{bmatrix} c & 0 \\ 0 & -qc \end{bmatrix}$, that is, it simply reverses the signs of the first negative diagonal entry and its successor. Obviously, only a finite number of such operations are performed—without ever reverting to an operation of type 4—before all diagonal entries with the possible exception of one in the last column are nonnegative.

Finally, if neither rule 4 nor rule 5 applies to a diagonal matrix, either the matrix is strongly diagonal, or the diagonal entry in the last column is negative and lies in a row other than the last row. In the latter case, rule 6 calls for subtraction of the row containing this entry from the row below. The transition to the next diagonal matrix then takes place in a 2×1 matrix where it is $\begin{bmatrix} -c \\ 0 \end{bmatrix} \sim \begin{bmatrix} -c \\ c \end{bmatrix} \sim \begin{bmatrix} 0 \\ c \end{bmatrix} \sim \begin{bmatrix} c \\ c \end{bmatrix} \sim \begin{bmatrix} c \\ 0 \end{bmatrix}$. In short, the sign of the last diagonal entry is simply reversed. This obviously does not create the need for any operations of type 4 or 5, and the algorithm terminates with a strongly diagonal matrix equivalent to the given one.

3 Equivalence of Strongly Diagonal Matrices

Theorem. *Two strongly diagonal matrices are equivalent only if they are equal.*

Proof. A $k \times k$ **minor** of a matrix (where k is a number that does not exceed either the number of rows or the number of columns of the matrix) is a number that is the determinant of a $k \times k$ matrix obtained by deleting all but k rows and all but k columns of the matrix. Given a matrix A, let $\mathcal{M}(A, k)$ denote the set of all $k \times k$ minors of A. For example, when $A = \begin{bmatrix} 4 & 0 & 2 & 4 \\ 2 & 2 & -2 & 4 \\ 1 & -2 & 5 & -1 \end{bmatrix}$, $\mathcal{M}(A, 2) = \{8, -12, -4, -8, 16, 18, 4, -22, -6, 12, 6, -18\}$ because the 2×2 minors of this matrix are the 18 determinants

$$\begin{vmatrix} 4 & 0 \\ 2 & 2 \end{vmatrix} = 8, \quad \begin{vmatrix} 4 & 2 \\ 2 & -2 \end{vmatrix} = -12, \quad \begin{vmatrix} 4 & 4 \\ 2 & 4 \end{vmatrix} = 8, \quad \begin{vmatrix} 0 & 2 \\ 2 & -2 \end{vmatrix} = -4, \ldots, \begin{vmatrix} -2 & 4 \\ 5 & -1 \end{vmatrix} = -18$$

which produce, when duplicates are eliminated, the above list of 12 integers. For this same A we have $\mathcal{M}(A, 1) = \{4, 0, 2, -2, 1, 5, -1\}$ and $\mathcal{M}(A, 3) = \{12, 0, -12\}$.

A $k \times k$ minor of TA, where T is a tilt, is either a $k \times k$ minor of A (if the row in which TA has a sum or difference of two rows of A is not one of the k chosen rows) or it is a $k \times k$ minor of A plus or minus another $k \times k$ determinant (otherwise); the second $k \times k$ determinant is either 0 (if the row that is added to or subtracted from another by multiplication on the left by T is one of the k chosen rows) or it is a $k \times k$ minor of A (otherwise). Therefore, every integer in the set $\mathcal{M}(TA, k)$ is either in the set $\mathcal{M}(A, k)$ or it is a sum or difference of two integers in this set. Therefore, any common divisor of the integers in $\mathcal{M}(A, k)$ divides all integers in $\mathcal{M}(TA, k)$. Similarly, any common divisor of the integers of $\mathcal{M}(A, k)$ divides all integers in $\mathcal{M}(AT, k)$. Therefore, if A and B are equivalent matrices, any common divisor of the integers in $\mathcal{M}(A, k)$ divides all integers of $\mathcal{M}(B, k)$. Moreover, $\mathcal{M}(A, k) = \{0\}$—that is, all $k \times k$ minors of A are zero—if and only if $\mathcal{M}(B, k) = \{0\}$.

Suppose now that D and E are equivalent strongly diagonal $m \times n$ matrices. Let r be the number of nonzero entries of D, and let s be the number of nonzero entries of E. Then $\mathcal{M}(D, r)$ contains a nonzero number (the product of the first r diagonal entries of D is a nonzero $r \times r$ minor of D) so $\mathcal{M}(E, r)$ contains a nonzero number. Therefore, $r \leq s$. By the same argument, $s \leq r$, so $r = s$.

Let $d_1, d_2, \ldots, d_r$ be the nonzero entries, in order, of D, and let $e_1, e_2, \ldots, e_r$ be the nonzero entries, in order, of E. For $k \le r$, $d_1 d_2 \cdots d_k$ divides all $k \times k$ minors of D. (A $k \times k$ minor of D which is not zero is a product of k diagonal entries of D and is therefore divisible by the product of the *first* k diagonal entries of D because D is strongly diagonal.) Therefore, $d_1 d_2 \cdots d_k$ divides $e_1 e_2 \cdots e_k$ (because $e_1 e_2 \cdots e_k$ is the determinant of the $k \times k$ matrix in the upper left corner of E). Since $e_1 e_2 \cdots e_k$ also divides $d_1 d_2 \cdots d_k$, it follows that $d_1 d_2 \cdots d_k = \pm e_1 e_2 \cdots e_k$. Except when $k = m = n$, the numbers $d_1, d_2, \ldots, d_k$ and $e_1, e_2, \ldots, e_k$ are all positive, so $d_1 d_2 \cdots d_k = e_1 e_2 \cdots e_k$ except in this case. When $k = m = n$, this equation holds because equivalent matrices have equal determinants. Thus, the product of the first k nonzero diagonal entries of D is always equal to the product of the first k nonzero diagonal entries of E, and division shows that each nonzero diagonal entry of D is equal to the corresponding diagonal entry of E, which implies $D = E$.

4 Conclusion

To determine whether two $m \times n$ matrices are equivalent, it suffices to apply the algorithm of Section 2 to find strongly diagonal matrices equivalent to each; the original matrices are equivalent if and only if the strongly diagonal matrices are equal.

5 The Rank of a Matrix

The **rank** of a matrix is the number of nonzero entries in the equivalent strongly diagonal matrix. As the proof just given shows, the rank of A can also be described as the largest integer k for which A contains a nonzero $k \times k$ minor. This observation implies that the rank of A is the number of nonzero entries in any diagonal matrix equivalent to A, not just the strongly diagonal matrix equivalent to A.

As was noted at the end of Chapter 3, a matrix A is a left zero divisor if and only if a diagonal matrix D equivalent to A has a column of zeros. It follows that *A is a left zero divisor if and only if its rank is less than the number of its columns.* Similarly, A is a right zero divisor if and only if its rank is less than the number of its rows.

Examples

The algorithm of Section 2 applied to $\begin{bmatrix} -6 & 0 \\ 0 & 4 \end{bmatrix}$ gives

$$\begin{bmatrix} -6 & 0 \\ 0 & 4 \end{bmatrix} \sim \begin{bmatrix} -6 & 4 \\ 0 & 4 \end{bmatrix} \sim \begin{bmatrix} -2 & 4 \\ 4 & 4 \end{bmatrix} \sim \begin{bmatrix} 2 & 4 \\ 8 & 4 \end{bmatrix} \sim \begin{bmatrix} 2 & 2 \\ 8 & -4 \end{bmatrix} \sim \begin{bmatrix} 2 & 0 \\ 8 & -12 \end{bmatrix} \sim \cdots \sim \begin{bmatrix} 2 & 0 \\ 0 & -12 \end{bmatrix}.$$

This equivalence is also clear on inspection, because 2 is obviously the greatest common divisor of the 1×1 minors of $\begin{bmatrix} -6 & 0 \\ 0 & 4 \end{bmatrix}$ and the determinant of this matrix is -24. Therefore, the first diagonal entry of the strongly diagonal matrix equivalent to this matrix must be 2 and the product of the two diagonal entries must be -24.

When just the diagonal matrices in the sequence generated by the algorithm of Section 2 are shown, the algorithm gives, for example,

$$\begin{bmatrix} -4 & 0 & 0 & 0 \\ 0 & -6 & 0 & 0 \\ 0 & 0 & -8 & 0 \\ 0 & 0 & 0 & 2 \end{bmatrix} \sim \begin{bmatrix} 2 & 0 & 0 & 0 \\ 0 & 12 & 0 & 0 \\ 0 & 0 & -8 & 0 \\ 0 & 0 & 0 & 2 \end{bmatrix} \sim \begin{bmatrix} 2 & 0 & 0 & 0 \\ 0 & 4 & 0 & 0 \\ 0 & 0 & -24 & 0 \\ 0 & 0 & 0 & 2 \end{bmatrix} \sim \begin{bmatrix} 2 & 0 & 0 & 0 \\ 0 & 4 & 0 & 0 \\ 0 & 0 & 2 & 0 \\ 0 & 0 & 0 & -24 \end{bmatrix} \sim \begin{bmatrix} 2 & 0 & 0 & 0 \\ 0 & 2 & 0 & 0 \\ 0 & 0 & 4 & 0 \\ 0 & 0 & 0 & -24 \end{bmatrix}.$$

If $A = \begin{bmatrix} 1 & 2 & 1 \\ 1 & 0 & 7 \\ 1 & 2 & 3 \end{bmatrix}$ then $\mathcal{M}(A, 1) = \{1, 2, 0, 7, 3\}$, $\mathcal{M}(A, 2) = \{-2, 6, 14, 0, 2, 4, -4, -14\}$, and $\mathcal{M}(A, 3) = \{-4\}$. From this information, it is easy to conclude that the equivalent strongly diagonal matrix is $\begin{bmatrix} 1 & 0 & 0 \\ 0 & 2 & 0 \\ 0 & 0 & -2 \end{bmatrix}$. Application of the algorithm of Section 2 to this A is easily seen to give this result (without, in fact, ever evoking rules 4, 5, or 6).

Exercises

1 Find strongly diagonal matrices equivalent to each of the following.

$$\text{(a)}\begin{bmatrix} 4 & 0 & 0 & 0 \\ 0 & -3 & 0 & 0 \\ 0 & 0 & -2 & 0 \\ 0 & 0 & 0 & -1 \end{bmatrix} \quad \text{(b)}\begin{bmatrix} -4 & 0 & 0 \\ 0 & -3 & 0 \\ 0 & 0 & -2 \\ 0 & 0 & 0 \end{bmatrix} \quad \text{(c)}\begin{bmatrix} -1 & 2 & 1 & 5 \\ -1 & 1 & 3 & -5 \\ -3 & 5 & 7 & -1 \\ -1 & 1 & 3 & -9 \end{bmatrix}$$

2 For the matrix $A = \begin{bmatrix} 1 & 2 \\ 3 & 4 \\ 5 & 6 \end{bmatrix}$ find $\mathcal{M}(A, 2)$. Based on this information and on the fact that 1 is an entry of A, find the strongly diagonal matrix equivalent to A. Give a sequence of linked matrices starting with A and ending with a strongly diagonal matrix and find, for each intermediate matrix B, $\mathcal{M}(B, 2)$.

3 Find the greatest common divisor of the $k \times k$ minors of $\begin{bmatrix} 2 & 5 & 4 \\ 1 & 1 & 2 \\ 1 & -2 & 2 \end{bmatrix}$ for $k = 1, 2, 3$. On the basis of this information, find the strongly diagonal matrix equivalent to this matrix.

4 Prove that a square matrix is unimodular if and only if its determinant is 1.

5 Find all strongly diagonal matrices that are equivalent to invertible matrices. (See Exercise 16 of Chapter 3.)

6 The following exercise shows *without using determinants* that if two strongly diagonal matrices are equivalent then each diagonal entry of one is an integer multiple of the corresponding diagonal entry of the other. It follows that they are equal unless they are square, the last diagonal entries are nonzero, and one is obtained from the other by reversing the sign of the last diagonal entry. The theory of Chapter 4 is needed to assure that this second possibility cannot occur, that is, that a strongly diagonal square matrix in which the last diagonal entry is not zero cannot be equivalent to the matrix obtained by reversing the sign of the last diagonal entry.

Let D and E be equivalent strongly diagonal $m \times n$ matrices, say $E = MDN$ where M and N are unimodular. Let $d_1, d_2, \ldots, d_k$ and $e_1, e_2, \ldots, e_k$ be the diagonal entries of D and E respectively, where k is the mininum of m and n. It is to be shown that e_i is an integer multiple of d_i for each i. Fill in the steps of the following proof.

Let C be the $(i-1) \times i$ matrix in the first $i-1$ rows of the first i columns of N. Let U and V be unimodular matrices such that UCV is strongly diagonal. The last column of CV is zero. Let W be the last column of V, and let $\hat{W}$ be the $n \times 1$ matrix whose first i entries are W and whose remaining entries (if any) are zero. The first $i-1$ entries of $N\hat{W}$ are all zero. All entries of $DN\hat{W}$ are multiples of d_i. All entries of $E\hat{W}$ are multiples of d_i. Each entry of $\hat{W}e_i$ is a multiple of the corresponding entry of $E\hat{W}$. Multiplication of the last column of V by e_i gives an $i \times 1$ matrix whose entries are multiples of d_i. Therefore e_i is a multiple of d_i, as was to be shown.

Supplementary Unit: Finitely Generated Abelian Groups

Let A be an $m \times n$ matrix of integers. Two $m \times 1$ matrices of integers a and b will be called **congruent mod** A, denoted $a \equiv b \bmod A$, if their difference is divisible on the left by A, that is, if there is an $n \times 1$ matrix c such that $a - b = Ac$. It is easy to see that congruence mod A for a fixed matrix A is an equivalence relation (reflexive, symmetric, and transitive).

If $a \equiv b \bmod A$ then $a + a' \equiv b + a' \bmod A$ for all $m \times 1$ matrices a'. Therefore, $a \equiv b \bmod A$ and $a' \equiv b' \bmod A$ imply $a + a' \equiv b + a' \equiv b + b' \bmod A$, so *congruence classes can be added* by adding representatives.

The set of all congruence classes of $m \times 1$ matrices mod A with this operation of addition is an *Abelian group.* This means that addition is associative and commutative, and that, given any two equivalence classes, say the classes represented by the $m \times 1$ matrices a and b, there is one and only one equivalence class whose addition to the class of a gives the class of b. (This class is, of course, the class represented by the matrix $b - a$. To see that this is the unique class whose addition to the class of a gives the class of b, note that its addition to the class of a gives, by definition, the class of $(b - a) + a = b$; if f is any matrix for which the class of $a + f$ is the class of b then $a + f \equiv b \bmod A$, which means that $(a + f) - b = f - (b - a)$ is of the form Ac for some $n \times 1$ matrix c, which is to say that $f \equiv b - a \bmod A$, that is, the class of f is the class of $b - a$.)

We will call the group defined in this way G_A.

For any element x of G_A, say the element represented by the matrix a, and for any integer k, $k \cdot x$ will denote the element of G_A represented by the matrix ak (a matrix product in which k is a 1×1 matrix). Clearly, $1 \cdot x = x$, $k \cdot x + l \cdot x = (k + l) \cdot x$, and $(kl) \cdot x = k \cdot (l \cdot x)$. If k is a positive integer, then $k \cdot x$ is the element of G_A obtained by adding x to itself k times. For any two elements x and y of G_A, $0 \cdot x = 0 \cdot y$, because both are the element of G_A represented by the $m \times 1$ matrix whose entries are all zero. This element of G_A will be denoted by 0. Clearly 0 is an *identity element* of G_A; that is, for any x in G_A the equation $x + 0 = x$ holds. Note that $(-1) \cdot x$ is the *additive inverse* of x in the sense that $(-1) \cdot x + x = (-1 + 1) \cdot x = 0 \cdot x = 0$. We will also write $-x$ for $(-1) \cdot x$.

1 Let $A = \begin{bmatrix} 2 & 0 \\ 0 & 3 \end{bmatrix}$. Show that G_A has 6 elements. Show, moreover, that there is an element g of G_A such that each element of G_A can be written in one and only one way in the form $k \cdot g$, where k is a positive integer $k \le 6$.

2 In the case $A = \begin{bmatrix} 3 & 0 \\ 0 & -6 \end{bmatrix}$, show that G_A contains 18 elements. More specifically, show that it contains two elements x and y such that (1) $3 \cdot x = 0$, (2) $6 \cdot y = 0$, and (3) every element of G_A can be written in one and only one way in the form $k \cdot x + l \cdot y$ where k and l are positive integers with $k \le 3$ and $l \le 6$.

3 Prove that the matrix $B = \begin{bmatrix} 3 & 6 \\ 9 & 12 \end{bmatrix}$ is equivalent to the matrix A of the preceding exercise. Conclude that G_B contains elements x' and y' with properties (1), (2), and (3) of the preceding exercise. In particular, G_B contains 18 elements.

4 Prove that if $A = \begin{bmatrix} 1 & 0 \\ 0 & 3 \\ 0 & 0 \end{bmatrix}$ then every 3×1 matrix a is congruent mod A to one and only one matrix of the form $\begin{bmatrix} 0 \\ b_2 \\ b_3 \end{bmatrix}$, where b_2 is 0, 1, or 2 and b_3 is an integer.

5 Let A be a diagonal matrix. Give necessary and sufficient conditions for G_A to contain a finite number of elements. When the number of elements is finite, describe it in terms of the entries of A.

6 For an arbitrary matrix A, how can one determine whether G_A contains a finite number of elements, and, in case it does, how can one find the number of elements it contains?

7 Two groups are said to be **isomorphic** if there is a one-to-one correspondence between their elements which "respects the addition operation" in the sense that if x and y correspond to x' and y' respectively, then $x + y$ corresponds to $x' + y'$. Prove that if A and B are equivalent then G_A and G_B are isomorphic.

8 The **order** of an element x of G_A is the smallest positive integer k such that $k \cdot x = 0$. (The order of 0 is 1, and an element has order ∞ if there is no positive integer k with $k \cdot x = 0$.) In the case $A = \begin{bmatrix} 6 & 0 & 0 \\ 0 & 10 & 0 \end{bmatrix}$, for what integers k does G_A contain an element of order k? Find all elements of order 2.

9 Let A be the 3×3 diagonal matrix with diagonal entries 6, 6, and 4. For what integers i does G_A contain an element of order i? Find all elements of order 2. Find all elements of order 3. Find an element of maximum order.

10 Let $A = \begin{bmatrix} 2 & 0 \\ 0 & 2 \end{bmatrix}$ and $B = \begin{bmatrix} 1 & 0 \\ 0 & 4 \end{bmatrix}$. Prove that G_A and G_B are not isomorphic.

11 Show that if A is strongly diagonal, G_A can be described as follows: There are integers $r \geq 0$ and $s \geq 0$, integers $c_1, c_2, \ldots, c_r$ greater than 1 with with property that c_i is a multiple of c_{i-1} for $i = 2, 3, \ldots, r$, and elements $x_1, x_2, \ldots, x_r, y_1, y_2, \ldots, y_s$ of G_A such that (1) $c_i x_i = 0$ for $i = 1, 2, \ldots, r$ and (2) each element of G_A can be written in one and only one way in the form $j_1 x_1 + j_2 x_2 + \cdots + j_r x_r + k_1 y_1 + k_2 y_2 + \cdots + k_s y_s$, where the js are integers satisfying $0 \leq j_i < c_i$ and the ks are integers. Be especially careful to describe the integer r.

12 Show that the *group* G_A determines the integers $r, s, c_1, c_2, \ldots, c_r$ in the description of G_A in the preceding exercise. In other words, if A and B are strongly diagonal, and if G_A and G_B are isomorphic groups, then the integers $r, s, c_1, c_2, \ldots, c_r$ are the same in the two cases.

13 Let A and B be given matrices. Describe a procedure for determining whether G_A and G_B are isomorphic.

14 Prove that $\begin{bmatrix} 1 & 0 \\ 0 & kl \end{bmatrix} \sim \begin{bmatrix} k & 0 \\ 0 & l \end{bmatrix}$ if and only if k and l are relatively prime, that is, if and only if their gcd is zero.

15 Show that, for any given A, there is a diagonal matrix B in which each diagonal entry is zero or a power of a prime, and which has the property that G_A is isomorphic to G_B.

16 Show that if A and B are both diagonal, if all diagonal entries of A and B are either zero or a power of a prime, and if G_A and G_B are isomorphic, then the diagonal entries of A and B are the same numbers, although they may appear in a different order.

17 (For readers familiar with the definition of an Abelian group.) Show that every finitely generated Abelian group is isomorphic to G_A for some matrix A.

Chapter 6
Matrices with Rational Number Entries

1 Introduction

IN THE DEFINITION of a linear substitution in Chapter 1, one can allow the multipliers to be rational numbers as well as integers. When this is done, the notion of a "linear substitution" is extended—what was called a linear substitution before is still a linear substitution, but new linear substitutions are allowed. The operation of **composition** of linear substitutions is defined in the same way as before, and it corresponds in the same way as before to an operation of **matrix multiplication** of matrices with rational entries.

The notion of **equivalence** of matrices with rational entries can be defined in the same way as in Chapter 2: Two matrices are **equivalent** if one can be obtained from the other by a sequence of operations in which one row or column is added to or subtracted from an adjacent row or column.

Given a matrix of rational numbers A and a number r, let rA denote* the matrix obtained by multiplying each entry of A by r. When $r \neq 0$, two matrices A and B are equivalent if and only if rA and rB are equivalent; this statement follows immediately from the definition because, for a tilt T, $B = AT$ or $B = TA$ if and only if $rB = (rA)T$ or $rB = T(rA)$, respectively. Given a matrix of rational numbers A, there is a nonzero integer r—a common denominator of the entries of A—for which rA is a matrix of integers. If M and N are unimodular matrices such that $M(rA)N = D$ is diagonal, then $MAN = r^{-1}D$ is a diagonal matrix of rational numbers equivalent to A. Moreover, the algorithm of Chapter 2 applies *without change* to the reduction of an arbitrary matrix of rational numbers to an equivalent diagonal matrix, because the steps of the algorithm do not in any way require that the entries of a matrix be integers. (The proof that the algorithm terminates does use the assumption that the entries are integers. When

*The notation rA, where r is a number and A is a matrix, always means the matrix obtained by multiplying each entry of A by r. Note that this is *not* a matrix multiplication unless A has just one row. It is sometimes helpful to think of the symbol r in rA as standing not for the number r but for the matrix rI_m, where m is the number of rows of A. With this interpretation of r in rA, it *is* a matrix multiplication. When A has just one column it is natural to write Ar in place of rA, because Ar is then a matrix multiplication.

the entries are rational numbers, the algorithm terminates because it takes exactly the same steps as does the algorithm applied to a matrix of integers obtained by multiplying by a common denominator of the entries.)

A **determinant** of an $n \times n$ matrix A whose entries are rational numbers is the product of the diagonal entries of a diagonal matrix equivalent to A. Let r be a common denominator of the entries of A. A diagonal matrix D is equivalent to A if and only if the diagonal matrix rD is equivalent to rA. The product of the diagonal entries of rD is r^n times the product of the diagonal entries of D, so a number is a determinant of rA if and only if it is r^n times a determinant of A. Because rA has only one determinant (Chapter 4), it follows that A has only one determinant, namely, $|rA|/r^n$. Naturally, this number will be denoted $|A|$.

When $|rA|$ is expanded as a sum of $n!$ terms, where each term is a product of n entries of rA, division by r^n shows that $|A|$ is a sum of $n!$ terms, where each term is a product of n entries of A. In short, the formula for the determinant of a matrix applies to matrices with rational entries as well as to matrices with integer entries.

The formula $|AB| = |A||B|$ holds for the determinant of a product of two matrices with rational entries, as follows either from a review of the proof in the case of matrices with integer entries to see that the same argument applies, or from the formula $|AB| = |rsAB|/(rs)^n = |rA||sB|/r^n s^n = |A||B|$, where r is a common denominator of the entries of A and s is a common denominator of the entries of B, so that rs is a common denominator of the entries of AB.*

A diagonal matrix of rational numbers is *strongly* diagonal if (1) each diagonal entry after the first is an integer multiple of the preceding diagonal entry and (2) there are no negative entries, except that the last diagonal entry of a strongly diagonal square matrix may be negative. Since, for any positive rational number r, $A \sim B$ if and only if $rA \sim rB$, and since A is strongly diagonal if and only if rA is, the theorem of the last chapter implies immediately that a matrix with rational entries is equivalent to one and only one strongly diagonal one.

2 *Matrix Division*

Let A and Y be matrices with the same number of rows. Chapter 3 showed how to find all matrices X with integer entries such that $AX = Y$. In Chapter 3, A and Y were assumed to have integer entries, but $AX = Y$ if and only if

*A third way to prove $|AB| = |A||B|$ is to regard it as a statement about two polynomials $|AB|$ and $|A||B|$ in $2n^2$ variables (the entries of the two $n \times n$ matrices). As was shown in Chapter 4, these two polynomials have the same values whenever the variables they contain are given integer values. It follows that they are equal *as polynomials* and therefore that they have equal values when the variables are given any numerical values whatsoever.

$rAX = rY$ (when $r \neq 0$), so the case in which A and Y have rational entries can be reduced to the case in which they have integer entries merely by finding a common denominator r of the entries of A and Y.

To find all matrices X with *rational* entries that solve the equation $AX = Y$ is a quite different problem, normally with more solutions. After all, the whole reason for computing with rational numbers is to have a system of numbers in which an equation of the form $ax = y$ always has a solution provided $a \neq 0$.

The method of Chapter 3 can easily be adapted to find all solutions X of $AX = Y$ with rational entries whenever A and Y are given matrices (rational entries) with the same number of rows. Exactly as in Chapter 3, the solutions X of $AX = Y$ with rational entries correspond one-to-one to solutions Z of $DZ = MY$ with rational entries via $X = NZ$, $Z = N^{-1}X$, where, of course, M and N are unimodular matrices such that $D = MAN$ is diagonal. Finding all solutions Z of $DZ = MY$ with rational entries, when MY and D are given and D is diagonal, is simple. There are no solutions Z of $DZ = MY$ unless MY is zero in all rows in which D is zero. When this condition is met, there are solutions, and the most general solution is a matrix Z which has in its ith row the ith row of MY divided by the ith diagonal entry of D whenever the ith diagonal entry of D is nonzero, and which may have arbitrary rational entries in the remaining rows. In short, the only changes are that rows of MY can now be divided by the corresponding diagonal entry of D provided only that the diagonal entry is *not zero,* and the arbitrary entries of Z are arbitrary rational numbers.

3 *Inverses*

The primary reason for using matrices with rational entries is to make possible the solution of equations $AX = Y$ in cases where there are no solutions in integers. In particular, the use of rational entries makes it possible to *invert* many matrices which have no inverses with integer entries.

Let A be an $m \times n$ matrix. A **right inverse** of A is a matrix B for which $AB = I$. Here I is necessarily the $m \times m$ identity matrix (it has the same number of rows as A), and B is necessarily $n \times m$ (it has the same number of columns as I_m and the same number of rows as A has columns). According to the solution of the problem of matrix division just given, A has a right inverse if and only if M is zero in all rows in which $MAN = D$ is zero. Since no row of M is zero (a zero row in M would imply a zero row in $MM^{-1} = I_m$) it follows that A has a right inverse if and only if each row of the diagonal matrix $MAN = D$ is nonzero. In other words, *a matrix has a right inverse if and only if its rank is equal to the number of its rows.* (For a matrix of rational numbers, as for a matrix of integers, the rank is the number of nonzero entries in the equivalent strongly diagonal matrix, or, what is the same, the largest integer k for which the matrix contains a nonzero $k \times k$ minor.) Note that A has a right inverse if and only if it is not a right zero divisor (see the last section of Chapter 5).

Similarly, a **left inverse** C of A is an $n \times m$ matrix for which $CA = I$. A matrix has a left inverse if and only if its rank is equal to the number of its columns, which is true if and only if it is not a left zero divisor.

A matrix is **invertible*** if it has both a right inverse and a left inverse. If A is invertible, if B is a right inverse of A, and if C is a left inverse of A, then $C = CI = CAB = IB = B$, so A has only one right inverse and only one left inverse, and they are the same matrix; this unique inverse of A on either side is called the **inverse** of A and is denoted A^{-1}. Since A is invertible if and only if its rank is equal to both the number of its rows and the number of its columns, *an invertible matrix must be square, and an $n \times n$ matrix is invertible if and only if its rank is n.*

If an $n \times n$ matrix has a right inverse, its rank must be n; therefore, it must be invertible, so it has only one right inverse, which is also the unique left inverse. Similarly, an $n \times n$ matrix which has a left inverse is necessarily invertible.

Examples

The algorithm of Chapter 2 applies without change to the matrix

$$A = \begin{bmatrix} \frac{1}{2} & \frac{1}{3} \\ \frac{1}{3} & \frac{1}{4} \end{bmatrix}$$

to give

$$\begin{bmatrix} \frac{1}{2} & \frac{1}{3} & 1 & 0 \\ \frac{1}{3} & \frac{1}{4} & 0 & 1 \\ 1 & 0 & 0 & 0 \\ 0 & 1 & 0 & 0 \end{bmatrix} \sim \begin{bmatrix} \frac{1}{6} & \frac{1}{3} & 1 & 0 \\ \frac{1}{12} & \frac{1}{4} & 0 & 1 \\ 1 & 0 & 0 & 0 \\ -1 & 1 & 0 & 0 \end{bmatrix} \sim \begin{bmatrix} \frac{1}{6} & \frac{1}{6} & 1 & 0 \\ \frac{1}{12} & \frac{1}{6} & 0 & 1 \\ 1 & -1 & 0 & 0 \\ -1 & 2 & 0 & 0 \end{bmatrix}$$

$$\sim \begin{bmatrix} \frac{1}{6} & 0 & 1 & 0 \\ \frac{1}{12} & \frac{1}{12} & 0 & 1 \\ 1 & -2 & 0 & 0 \\ -1 & 3 & 0 & 0 \end{bmatrix} \sim \begin{bmatrix} \frac{1}{12} & -\frac{1}{12} & 1 & -1 \\ \frac{1}{12} & \frac{1}{12} & 0 & 1 \\ 1 & -2 & 0 & 0 \\ -1 & 3 & 0 & 0 \end{bmatrix}$$

$$\sim \begin{bmatrix} \frac{1}{12} & 0 & 1 & -1 \\ \frac{1}{12} & \frac{1}{6} & 0 & 1 \\ 1 & -1 & 0 & 0 \\ -1 & 2 & 0 & 0 \end{bmatrix} \sim \begin{bmatrix} \frac{1}{12} & 0 & 1 & -1 \\ 0 & \frac{1}{6} & -1 & 2 \\ 1 & -1 & 0 & 0 \\ -1 & 2 & 0 & 0 \end{bmatrix},$$

*This definition supersedes the definition of earlier chapters, in which "invertible" meant that the inverse had integer entries.

which results in the explicit equivalence

$$\begin{bmatrix} 1 & -1 \\ -1 & 2 \end{bmatrix}\begin{bmatrix} \frac{1}{2} & \frac{1}{3} \\ \frac{1}{3} & \frac{1}{4} \end{bmatrix}\begin{bmatrix} 1 & -1 \\ -1 & 2 \end{bmatrix} = \begin{bmatrix} \frac{1}{12} & 0 \\ 0 & \frac{1}{6} \end{bmatrix}.$$

Alternatively, one can write $A = \frac{1}{12}\begin{bmatrix} 6 & 4 \\ 4 & 3 \end{bmatrix}$, apply the algorithm of Chapter 2 to $\begin{bmatrix} 6 & 4 \\ 4 & 3 \end{bmatrix}$ to find $\begin{bmatrix} 1 & -1 \\ -1 & 2 \end{bmatrix}\begin{bmatrix} 6 & 4 \\ 4 & 3 \end{bmatrix}\begin{bmatrix} 1 & -1 \\ -1 & 2 \end{bmatrix} = \begin{bmatrix} 1 & 0 \\ 0 & 2 \end{bmatrix}$, and divide by 12 to find the same equivalence.

When solutions X of $AX = Y$ with rational entries are allowed, the first example $A = \begin{bmatrix} 1 & 0 & 0 \\ 0 & 3 & 0 \end{bmatrix}$, $Y = \begin{bmatrix} 1 & 2 \\ 3 & 4 \end{bmatrix}$ of Chapter 3 has the solution $X = \begin{bmatrix} 1 & 2 \\ 1 & \frac{4}{3} \\ a & b \end{bmatrix}$, where a and b are arbitrary rational numbers. The solution of the second example, in which A is the same and Y is $\begin{bmatrix} 1 & 2 \\ -3 & 9 \end{bmatrix}$, is the solution $X = \begin{bmatrix} 1 & 2 \\ -1 & 3 \\ a & b \end{bmatrix}$ given in Chapter 3, except that now a and b are arbitrary rational numbers instead of arbitrary integers.

Allowing solutions X with rational entries does not change the solutions of the next two examples of Chapter 3; that is, for $A = \begin{bmatrix} 1 & 0 \\ 0 & 3 \\ 0 & 0 \end{bmatrix}$ there is still no solution when $Y = \begin{bmatrix} 1 \\ 3 \\ 5 \end{bmatrix}$, and there is only the solution $X = \begin{bmatrix} 2 \\ 2 \end{bmatrix}$ when $Y = \begin{bmatrix} 2 \\ 6 \\ 0 \end{bmatrix}$.

Exercises

1 Find the strongly diagonal matrix equivalent to

$$\text{(a)} \begin{bmatrix} \frac{1}{2} & \frac{3}{2} & \frac{1}{2} \\ 1 & 1 & 2 \\ \frac{3}{2} & \frac{3}{2} & \frac{5}{2} \end{bmatrix} \qquad \text{(b)} \begin{bmatrix} \frac{1}{3} & 0 & 1 \\ 3 & 1 & 10 \\ 1 & -1 & 5 \end{bmatrix}.$$

2 Find the most general right inverse of $A = \begin{bmatrix} 1 & 2 & 3 \\ 4 & 5 & 6 \end{bmatrix}$.

3 Find the most general right inverse of $A = \begin{bmatrix} 2 & 3 & 1 \\ 5 & 7 & 2 \end{bmatrix}$. What is the most general right inverse with integer entries? What left inverses does this A have?

4 Show that if A and B are equivalent matrices with rational entries and if A has integer entries, then B has integer entries.

5 As is noted in Section 3, a matrix has a left inverse if and only if it is not a left zero divisor, because both are true if and only if the rank is the number of columns. Give a simple proof that a matrix which has a left inverse is not a left zero divisor.

6 Let A be a square matrix with rational entries. Show that if B is obtained by adding a nonzero rational multiple of one row to another row then $|B| = |A|$.

7 Two matrices A and B with rational entries are **equivalent in the wider sense** if one can be obtained from the other by a succession of operations in which a rational multiple of one row is added to an adjacent row or a rational multiple of one column is added to an adjacent column (see Section 8 of Chapter 8). Prove that, unless $m = n = r$, any two $m \times n$ matrices of rank r are equivalent in the wider sense. What additional condition must be satisfied in the case $m = n = r$?

8 Show that if A and B are invertible $n \times n$ matrices, then AB is invertible and its inverse is $B^{-1}A^{-1}$.

9 Let A be a square matrix, and let M and N be unimodular matrices such that $MAN = D$ is diagonal. Show that if $|A| \neq 0$ then A is invertible and $A^{-1} = ND^{-1}M$, where D^{-1} is the diagonal matrix in which each diagonal entry is the reciprocal of the corresponding entry of D.

10 Use the method of the preceding exercise to find the inverse of $\begin{bmatrix} 1 & 4 & 1 \\ 2 & 10 & 5 \\ 1 & 4 & 2 \end{bmatrix}$.

11 Let D be a diagonal $m \times n$ matrix. Describe how to determine whether D has a left inverse, a right inverse, or neither, and describe how to find a left or right inverse of D when one exists.

12 If M and N are unimodular matrices such that $MAN = D$ is diagonal, and if D has a left inverse, show that A has a left inverse. Give a formula for the left inverse of A in this case. Answer the analogous questions for right inverses.

13 Let A be an invertible square matrix with integer entries. Find a simple way to test whether the entries of A^{-1} are all integers.

14 Show that a wide matrix need not have a right inverse, but that if a wide matrix has a right inverse it has more than one.

15 What changes, if any, are needed in the answer to Exercise 13 of Chapter 4 when it is used to find the inverse of an invertible matrix of rational numbers?

Supplementary Unit: Sets and Vector Spaces

Let $\mathbf{Q}^n$ denote the set of all $n \times 1$ matrices with rational entries. In this unit, such matrices will be denoted by boldface letters $\mathbf{u}$, $\mathbf{v}$, $\mathbf{w}$, and the set-theoretic notation $\mathbf{u} \in \mathbf{Q}^n$—in words, $\mathbf{u}$ is an element of the set $\mathbf{Q}^n$—will be used to mean "$\mathbf{u}$ is an $n \times 1$ matrix with rational entries."

If $\mathbf{u} \in \mathbf{Q}^n$ and if A is an $m \times n$ matrix, then the product matrix $A\mathbf{u}$ has m rows and just one column, that is, $A\mathbf{u} \in \mathbf{Q}^m$. Thus, multiplication on the left by A defines a *function* from $\mathbf{Q}^n$ to $\mathbf{Q}^m$.

1 For what $m \times n$ matrices A is the function $\mathbf{Q}^n \to \mathbf{Q}^m$ given by multiplication on the left by A a one-to-one function? That is, when is it true that $A\mathbf{u} = A\mathbf{v}$ implies $\mathbf{u} = \mathbf{v}$?

2 For what $m \times n$ matrices A is the function $\mathbf{Q}^n \to \mathbf{Q}^m$ given by multiplication on the left by A an onto function? That is, when is it true that each element of $\mathbf{Q}^m$ is of the form $A\mathbf{u}$ for some $\mathbf{u} \in \mathbf{Q}^n$?

3 Let S, T, and U be sets, and let $f : S \to T$ and $g : T \to U$ be functions from S to T and from T to U, respectively. The **composed function** of f and g, denoted $g \circ f$ (or, according to the convention adopted by some authors, $f \circ g$), is the function $S \to U$ which assigns to an element s of S the element $g\big(f(s)\big)$ of U. If A and B are matrices, when can the functions defined by multiplication on the left by A and B be composed, and what matrix corresponds to the composed function?

4 A **left inverse** of a function $f : S \to T$ is a function $g : T \to S$ for which the composed function $g \circ f$ is the identity function on the set S. Show that f has a left inverse if and only if it is one-to-one. (Assume S is nonempty; otherwise, the statement to be proved is difficult to interpret.)

5 A **right inverse** of a function $f : S \to T$ is a function $g : T \to S$ for which the composed function $f \circ g$ is the identity function on the set T. Show that f has a right inverse if and only if it is onto. (Again, assume S is nonempty, so the existence of an f implies T is nonempty.)

Thus, a function $f : \mathbf{Q}^n \to \mathbf{Q}^m$ has a left inverse if and only if it is one-to-one. When such a function is given by multiplication on the left by an $m \times n$ matrix A, it has this property if and only if $\operatorname{rank}(A) = n$, or, what is the same, if and only if A is not a left zero divisor. Similarly, such a function has a right inverse if and only if it is onto; this is true if and only if $\operatorname{rank}(A) = m$, or, what is the same, if and only if A is not a right zero divisor.

Matrix addition defines an operation of **addition** on $\mathbf{Q}^n$; that is, if $\mathbf{u}, \mathbf{v} \in \mathbf{Q}^n$ then $\mathbf{u} + \mathbf{v}$ is another element of $\mathbf{Q}^n$. Matrix multiplication defines an operation of **scalar multiplication** of elements of $\mathbf{Q}^n$ in which, given $\mathbf{u} \in \mathbf{Q}^n$ and $a \in \mathbf{Q}$, an element $\mathbf{u}a \in \mathbf{Q}^n$ is defined, namely, the element obtained from $\mathbf{u}$ by multiplying all entries by a.

The operations of addition and scalar multiplication have certain obvious properties of commutativity, associativity, and distributivity. Formally, the properties to be singled out are:

$$\mathbf{u} + \mathbf{v} = \mathbf{v} + \mathbf{u}, \quad (\mathbf{u} + \mathbf{v}) + \mathbf{w} = \mathbf{u} + (\mathbf{v} + \mathbf{w}), \quad \mathbf{u}(ab) = (\mathbf{u}a)b,$$
$$\mathbf{u}(a + b) = \mathbf{u}a + \mathbf{u}b, \quad (\mathbf{u} + \mathbf{v})a = \mathbf{u}a + \mathbf{v}a, \quad \mathbf{u}1 = \mathbf{u}, \quad \mathbf{u} + \mathbf{v}0 = \mathbf{u}.$$

A **vector space** is a set $\mathbf{V}$ on which there are given operations of addition and scalar multiplication by rational numbers such that the properties enumerated above hold. The reader may well be acquainted with the use of the word "vector" to mean "an arrow." Loosely speaking, the set of arrows in three-dimensional space can be thought of as a vector space; two arrows are added by putting the tail of one arrow at the head of the other and drawing the arrow which joins the free tail to the free head, while an arrow is multiplied by a rational number by taking the arrow with the same direction but with its length multiplied by the rational number (where multiplication by a negative number reverses the direction). Somewhat more precisely, the vector space $\mathbf{Q}^3$ can be visualized as the set of all arrows in xyz-space, where the three entries of an element of $\mathbf{Q}^3$ give the x-, y-, and z-components of the corresponding arrow.

6 Let $\mathbf{V}$ and $\mathbf{W}$ be vector spaces. A function $f : \mathbf{V} \to \mathbf{W}$ is said to be a **linear function** if it respects the vector space operations of addition and scalar multiplication, that is, if $f(\mathbf{v}_1+\mathbf{v}_2) = f(\mathbf{v}_1)+f(\mathbf{v}_2)$ for any pair of elements $\mathbf{v}_1, \mathbf{v}_2 \in \mathbf{V}$, and if $f(\mathbf{v}a) = f(\mathbf{v})a$ for any $\mathbf{v} \in \mathbf{V}$ and $a \in \mathbf{Q}$. Prove that a function $f : \mathbf{Q}^n \to \mathbf{Q}^m$ is linear if and only if there is an $m \times n$ matrix of rational numbers A such that f is the operation of multiplication on the left by A.

7 A set of elements $\mathbf{v}_1, \mathbf{v}_2, \ldots, \mathbf{v}_k$ of a vector space $\mathbf{V}$ gives rise to a function $\mathbf{Q}^k \to \mathbf{V}$, namely, the function that carries the element of $\mathbf{Q}^k$ whose entries are $a_1, a_2, \ldots, a_k$, in that order, to the element $\mathbf{v}_1a_1 + \mathbf{v}_2a_2 + \cdots + \mathbf{v}_ka_k$. The set $\mathbf{v}_1, \mathbf{v}_2, \ldots, \mathbf{v}_k$ is said to be **linearly independent** if this linear function $\mathbf{Q}^k \to \mathbf{V}$ is one-to-one. Show that the existence of a set of k linearly independent elements of $\mathbf{Q}^n$ implies $k \le n$.

8 A set $\mathbf{v}_1, \mathbf{v}_2, \ldots, \mathbf{v}_k$ of elements of a vector space $\mathbf{V}$ is said to **span** $\mathbf{V}$ if the linear function $\mathbf{Q}^k \to \mathbf{V}$ to which it gives rise is onto. Show that the existence of a set of k elements spanning $\mathbf{Q}^n$ implies $k \ge n$.

9 A set of elements $\mathbf{v}_1, \mathbf{v}_2, \ldots, \mathbf{v}_k$ of a vector space $\mathbf{V}$ is said to be a **basis of V** if they are linearly independent and span $\mathbf{V}$. A vector space $\mathbf{V}$ is said to be ***n*-dimensional** if it has a basis containing n elements. Prove that $\mathbf{Q}^n$ is n-dimensional.

10 A vector space $\mathbf{V}$ is said to be **finite-dimensional** if it is n-dimensional for some n. Prove that a finite-dimensional vector space has a well-defined dimension, that is, prove that if $\mathbf{V}$ is both n-dimensional and m-dimensional then $n = m$.

11 Let $\mathbf{V}$ and $\mathbf{W}$ be finite-dimensional vector spaces. Show that the existence of an onto linear function $\mathbf{V} \to \mathbf{W}$ implies that the dimension of $\mathbf{V}$ is at least as great as the dimension of $\mathbf{W}$, and that the existence of a one-to-one linear function $\mathbf{V} \to \mathbf{W}$ implies that the dimension of $\mathbf{V}$ is no greater than the dimension of $\mathbf{W}$. Note that these statements make "geometrical" sense: A low-dimensional space cannot be mapped onto a high-dimensional one. A high-dimensional space cannot be mapped one-to-one to a low-dimensional one.

Chapter 7

The Method of Least Squares

1 Introduction

AS WAS SEEN IN CHAPTER 3, an equation of the form $AX = Y$, in which A and Y are matrices with the same number of rows, may have have a unique solution X, but may also have many solutions or none at all. The method of least squares is a method of choosing a "best" solution in cases where there are many, and a method of choosing a "best approximation" to a solution when there is no actual solution. As will be shown in this chapter, there is a matrix B associated with A—we will call it the **mate** of A—with the property that $X = BY$ is the best choice for a solution of $AX = Y$ (even when there is no solution) in a very specific sense. The mate of A is defined in Section 3. First, the notion of the transpose of a matrix and a few properties of transposes need to be explained.

2 The Transpose of a Matrix

The **transpose** of an $m \times n$ matrix is the $n \times m$ matrix obtained by writing the rows as columns (and consequently the columns as rows). For example, the transpose of

$$\begin{bmatrix} 1 & 2 & 3 \\ 4 & 5 & 6 \end{bmatrix} \quad \text{is} \quad \begin{bmatrix} 1 & 4 \\ 2 & 5 \\ 3 & 6 \end{bmatrix}.$$

The transpose of A will be denoted* A^T. Clearly, the transpose of A^T is again A, that is, $(A^T)^T = A$. Somewhat less obvious, but not difficult to see, is the identity $(AB)^T = B^T A^T$, that is, if A and B are two matrices such that AB is defined (the number of columns of A is equal to the number of rows of B), then $B^T A^T$ is defined and equal to $(AB)^T$. (If A has just one row and B has just one column, then $AB = B^T A^T$ because both are equal to the sum of the products of corresponding entries of A and B. Let A_i be the ith row of A and B_j be the jth column of B. The entry in the ith row of the jth column of AB

*Note that T is *not an exponent* here. Note also that an exponent on a matrix can have meaning only when the matrix is square, so the danger of misinterpreting the superscript T as an exponent is not great.

is $A_i B_j = B_j^T A_i^T$, which is the entry in the jth row of the ith column of $B^T A^T$. That is, $AB = (B^T A^T)^T$, which implies $(AB)^T = B^T A^T$.)

A matrix is said to be **symmetric** if it is its own transpose. A symmetric matrix must, clearly, be square. A square matrix is symmetric if and only if every entry above the diagonal is equal to a corresponding entry below the diagonal; specifically, an $n \times n$ matrix is symmetric if and only if, for each pair of integers i and j satisfying $1 \leq i < j \leq n$, the entry in the ith row of the jth column is the same as the entry in the jth row of the ith column.

If A is any matrix, then $A^T A$ is defined (the number of columns of A^T is the number of rows of A) and symmetric ($(A^T A)^T = A^T (A^T)^T = A^T A$). Application of this observation to A^T instead of A shows that AA^T is also defined and symmetric for any matrix A.

Note that *if a symmetric matrix is invertible, its inverse is also symmetric.* This follows simply from the observation that if $A = A^T$ and A is invertible then $A(A^{-1})^T = A^T (A^{-1})^T = (A^{-1} A)^T = I^T = I$; thus $(A^{-1})^T$ is a right inverse of A, which implies $(A^{-1})^T = A^{-1}$.

3 *Mates*

Let A be a given $m \times n$ matrix. An $n \times m$ matrix B will be called a **mate** of A if it satisfies four conditions: $ABA = A$, $BAB = B$, $(AB)^T = AB$, and $(BA)^T = BA$. Note that the symmetry of these conditions implies that if B is a mate of A then A is a mate of B. Moreover, if A and B are mates, then A^T and B^T are mates, because then $A^T B^T A^T = (ABA)^T$ (by the rule for the transpose of a product) $= A^T$ (because A and B are mates), $(A^T B^T)^T = BA$ (by the rule for the transpose of a product) $= (BA)^T$ (because A and B are mates) $= A^T B^T$, and, by symmetry, $B^T A^T B^T = B^T$ and $(B^T A^T)^T = B^T A^T$.

The term "mate" is here introduced to describe what is generally called the "pseudo-inverse," the "generalized inverse," or the "Moore–Penrose inverse" of a matrix. As these terms suggest, the mate of a matrix is akin to an inverse. Indeed, if A is invertible, then A^{-1} is a mate of A. (When $B = A^{-1}$, both $AB = I$ and $BA = I$ are symmetric and $ABA = IA = A$ and $BAB = BI = B$ both hold.) Furthermore, as will be seen in Section 4, the mate of A is in a very natural sense the nearest thing to an inverse of A for any matrix A, even one which is not square and can therefore have no inverse.

Throughout this chapter, the matrices considered will be matrices whose entries are rational numbers.

Theorem. *Each matrix has one and only one mate.*

The proof uses:

Lemma. *If A is a matrix and $A^TA = 0$ then $A = 0$.*

Proof of the Lemma. The ith diagonal entry of A^TA is, by the definition of A^T and the definition of matrix multiplication, the sum of the squares of the entries in the ith column of A. Since a sum of squares of rational numbers is zero only if the numbers are all zero, the ith diagonal entry of A^TA is zero only if the ith column of A contains only zeros. If all entries of A^TA are zero, all entries of A must therefore be zero.

Proof of the Theorem. It will be shown first that no matrix has more than one mate. If B is a mate of A then $A^TAB = A^T(AB)^T$ (by the definition of a mate) $= (ABA)^T$ (by the rule for the transpose of a product) $= A^T$ (by the definition of a mate). Therefore, if both B and B_1 are mates of A, and if $C = B_1 - B$, it follows that $(AC)^TAC = C^TA^TAC = C^TA^TAB_1 - C^TA^TAB = C^TA^T - C^TA^T = 0$; by the Lemma, it follows that $AC = 0$. Thus $A(B_1 - B) = 0$, so $AB_1 = AB$. Since B_1^T and B^T are both mates of A^T, it follows that $A^TB_1^T = A^TB^T$. Taking the transpose of this equation then gives $B_1A = BA$. Thus, if both B and B_1 are mates of A, then $B_1 = B_1AB_1 = B_1AB = BAB = B$. In short, A has at most one mate.

If A has a left inverse, say $EA = I$, a mate of A can be found in the following way. The square matrix $S = A^TA$ is not a zero divisor, because if D is a matrix for which $SD = 0$, then $(AD)^T(AD) = D^TA^TAD = D^TSD = 0$ and $AD = 0$ by the Lemma; therefore, $D = EAD = E0 = 0$, so S is not a zero divisor. By Proposition 2 of Section 3.5, the determinant of S is not zero, so $S = A^TA$ is invertible (Section 3 of Chapter 6). Let $B = (A^TA)^{-1}A^T$. Then $BA = (A^TA)^{-1}A^TA = I$, so BA is symmetric, $BAB = B$, and $ABA = A$. Finally, $AB = A(A^TA)^{-1}A^T$ is symmetric, because $(A^TA)^{-1}$, being the inverse of a symmetric matrix, is symmetric, so $(AB)^T = B^TA^T = (A^T)^T(A^TA)^{-1}A^T = AB$. In summary, if A has a left inverse, then A^TA is invertible and $(A^TA)^{-1}A^T$ is the mate of A. Note that the mate of A is a left inverse of A.

Consider next the case in which A has a right inverse. Then A^T has a left inverse (namely, the transpose of a right inverse of A), so A^T has the mate $(AA^T)^{-1}A$, and this matrix is a left inverse of A^T. It follows that the transpose $A^T(AA^T)^{-1}$ of this matrix is the mate of A and a right inverse of A.

Suppose next that $A = FG$, where F has a left inverse and G has a right inverse. By what has already been shown, F and G both have mates, call them F^M and G^M respectively; moreover, $F^MF = I$ and $GG^M = I$. Set $B = G^MF^M$. Then $ABA = FGG^MF^MFG = FIIG = FG = A$, $BAB = G^MF^MFGG^MF^M = G^MF^M = B$, $BA = G^MF^MFG = G^MG =$ symmetric, and $AB = FGG^MF^M = FF^M =$ symmetric. Thus B is the mate of A.

Finally, for an arbitrary matrix A, there exist unimodular matrices M and N and a strongly diagonal matrix D such that $MAN = D$. Then $A = M^{-1}DN^{-1}$. Let r be the number of nonzero diagonal entries of D, let P be the $m \times r$ matrix whose columns are the first r columns of M^{-1}, let Q be the $r \times r$ matrix in the upper left corner of D, and let R be the $r \times n$ matrix whose rows are the first r rows of N^{-1}. Directly from the definition of matrix multiplication, it follows that $A = M^{-1}DN^{-1} = PQR$. (The first r columns of $M^{-1}D$ are PQ, and the remaining $n - r$ columns are zero. Therefore, the entry in row i and column j of $M^{-1}DN^{-1}$ is found by the same computation as the entry in row i and column j of $(PQ)R$, except that $n - r$ terms, all equal to zero, are omitted in the computation of the entry of $(PQ)R$. Thus, $M^{-1}DN^{-1} = PQR$.) Because $MM^{-1} = I$, the first r rows of M are a left inverse of P. Similarly, the first r columns of N are a right inverse of R. Finally, Q is invertible because it is a square matrix whose determinant is nonzero. By what was just shown, PQ has the mate $Q^M P^M$. Moreover, $Q^M P^M PQ = Q^M IQ = I$, so PQ has a left inverse. Therefore, $A = (PQ)R$ has a mate, and the proof is complete.

The proof gives an algorithm for finding the mate of A. If A is $m \times n$ and if r is its rank, there are three cases. If $r = n$, the mate of A is given by the formula $(A^TA)^{-1}A^T$. If $r = m$, it is given by the formula $A^T(AA^T)^{-1}$. Otherwise, let $MAN = D$ where M and N are unimodular and D is strongly diagonal. Let P be the first r columns of M^{-1}, let Q be the upper left $r \times r$ corner of D, and let R be the first r rows of N^{-1}. Then the mate of A is $R^M Q^M P^M$, where $R^M = R^T(RR^T)^{-1}$, $Q^M = Q^{-1}$, and $P^M = (P^TP)^{-1}P^T$. (Q^{-1} is *a* mate of Q, so it is *the* mate of Q.)

4 *Mates as Generalized Inverses*

Notation: Given a matrix A, the sum of the squares of the entries of A will be denoted $\|A\|^2$. It is natural to regard $\|A\|^2$ as a measure of how large A is, and therefore to regard $\|A - B\|^2$ as a measure of the amount by which two matrices A and B differ. (Here A and B are of course assumed to be matrices of the same size; otherwise $A - B$ would be meaningless.)

The problem of matrix division is the problem of finding all matrices X for which $AX = Y$, where A and Y are given matrices with the same number of rows. As was seen in Chapter 3, there may be *no* X satisfying $AX = Y$, or there may be *many*. If there are no solutions, it is natural to try to make AX as "near" to Y as possible, that is, to minimize $\|AX - Y\|^2$. We will say that X_0 is an **approximate solution** of $AX = Y$ if it has the property that $\|AX_0 - Y\|^2 \leq \|AX - Y\|^2$ for all matrices X. (It is assumed, of course, that AX_0 and AX are defined and have the same size as Y, that is, X_0 and X have the same number of rows as A has columns and the same number of columns as Y. Note that if $AX = Y$ has an actual solution then X_0 is an approximate solution if and only if it is an actual solution because $\|AX_0 - Y\|^2 \leq 0$ if and only if $AX_0 = Y$.) We will say that X_0 is a **best** approximate solution of $AX = Y$ if $\|X_0\|^2 \leq \|X\|^2$ for all approximate solutions X. With this terminology, the following theorem states simply that $AX = Y$ always has a unique best approximate solution and that it is given by the formula $X_0 = BY$, where B is the mate of A.

Theorem. *Let A and Y be matrices with the same number of rows, let B be the mate of A, and let $X_0 = BY$. For any matrix X of the same size as X_0, $\|AX - Y\|^2 \geq \|AX_0 - Y\|^2$. If $\|AX - Y\|^2 = \|AX_0 - Y\|^2$ and $X \neq X_0$, then $\|X\|^2 > \|X_0\|^2$.*

Proof. Let A have m rows and n columns. Note first that $(AB)^2 = ABAB = AB$ (because $ABA = A$) and $(AB)^T = AB$. Therefore, if U is any $m \times 1$ matrix, ABU is an $m \times 1$ matrix and $\|ABU\|^2 = (ABU)^T(ABU) = U^T(AB)^T ABU = U^T(AB)^2 U = U^T ABU$. Moreover, $(I - AB)^2 = (I - AB)I - (I - AB)AB = I - AB - AB + (AB)^2 = I - AB$ and $(I - AB)^T = I - (AB)^T = I - AB$; therefore, in a similar way, for any $m \times 1$ matrix U, $\|(I - AB)U\|^2 = U^T(I - AB)^T(I - AB)U = U^T(I - AB)^2 U = U^T(I - AB)U$. Thus, $\|ABU\|^2 + \|(I - AB)U\|^2 = U^T ABU + U^T(I - AB)U = U^T(AB + I - AB)U = U^T IU = \|U\|^2$. If U is any matrix with m rows, and if U_i is any column of U, then, as has just been shown, $\|ABU_i\|^2 + \|(I - AB)U_i\|^2 = \|U_i\|^2$. Since the sum of the squares of all elements of a matrix is equal to the sum over all columns of the sum of the squares of the entries in the column, summation of the identity $\|ABU_i\|^2 + \|(I - AB)U_i\|^2 = \|U_i\|^2$ over all columns of U gives the identity $\|ABU\|^2 + \|(I - AB)U\|^2 = \|U\|^2$ for any matrix U with m rows.* By the symmetry of the mate relation, it follows that $\|BAU\|^2 + \|(I - BA)U\|^2 = \|U\|^2$ for any matrix U with n rows.

Application of the first of these identities to the matrix $U = AX - Y$ where X is a matrix of the same size as X_0 gives $\|AX - Y\|^2 = \|AB(AX - Y)\|^2 + \|(I - AB)(AX - Y)\|^2 \geq \|(I - AB)(AX - Y)\|^2 = \|AX - Y - ABAX + ABY\|^2 = \|AX_0 - Y\|^2$, which is the first statement to be proved.

The inequality in the preceding paragraph shows, moreover, that if $\|AX - Y\|^2 = \|AX_0 - Y\|^2$ then $\|AB(AX - Y)\|^2 = 0$, that is, $AB(AX - Y) = 0$, which is to say $AX = ABY$. When X satisfies this condition, it follows that $\|X\|^2 = \|BAX\|^2 + \|(I - BA)X\|^2$ is at least as great as $\|BAX\|^2 = \|BABY\|^2 =$

*In the terminology of Chapter 10, the identity $\|ABU\|^2 + \|I - AB)U\|^2 = \|U\|^2$ expresses the fact that $AB + (I - AB) = I$ is an orthogonal partition of unity, a point of view that clarifies the calculations here.

$\|BY\|^2 = \|X_0\|^2$, and equality holds only when $\|(I - BA)X\|^2 = 0$, that is, only when $X = BAX = BABY = X_0$, which is the second statement to be proved.

This theorem implies that if A and Y have the same number of *columns,* then $X = YB$ is the best approximate solution of the division problem $XA = Y$ in the same sense: $\|XA - Y\|^2$ assumes its minimum value when $X = YB$, and, among all matrices X for which $\|XA - Y\|^2$ assumes its minimum value, YB is "smallest" in the sense that $\|YB\|^2$ is smallest. These statements follow immediately from the theorem just proved when one observes that $XA - Y$ is the transpose of $A^TX^T - Y^T$, that the sum of the squares of the entries of a matrix is equal to the sum of the squares of the entries of its transpose, and that the mate of A^T is B^T.

If A is an $m \times n$ matrix of rank n and Y is a matrix with m rows, the above method chooses $(A^TA)^{-1}A^TY$ as the "best" approximate solution of $AX = Y$. Because of the criterion by which the approximate solution was chosen, this method of "solving" $AX = Y$ is traditionally called the **method of least squares**.

Examples

The "best" approximate solution of $AX = Y$ when $A = \begin{bmatrix} 3 \\ 2 \end{bmatrix}$ and $Y = \begin{bmatrix} 2 \\ 1 \end{bmatrix}$ is found by finding the mate $B = (A^TA)^{-1}A^T = (3^2+2^2)^{-1}[\,3 \quad 2\,] = [\,\frac{3}{13} \quad \frac{2}{13}\,]$ of A and computing $BY = [\,\frac{3}{13} \quad \frac{2}{13}\,]\begin{bmatrix} 2 \\ 1 \end{bmatrix} = \frac{8}{13}$. In other words, $\frac{8}{13}$ times A, which is $\begin{bmatrix} \frac{24}{13} \\ \frac{16}{13} \end{bmatrix}$, is the multiple of A nearest Y, where "nearest" means that the sum of the squares of the differences of the coordinates is a minimum. Geometrically, this is the statement that the foot of the perpendicular from the point (2, 1) to the line through (0, 0) and (3, 2) is at the point $(\frac{24}{13}, \frac{16}{13})$.

In the case of this A,

$$AB = \begin{bmatrix} \frac{9}{13} & \frac{6}{13} \\ \frac{6}{13} & \frac{4}{13} \end{bmatrix}.$$

The identity $(AB)^2 = AB$ noted in Section 4 is verified by the calculation

$$\begin{bmatrix} \frac{9}{13} & \frac{6}{13} \\ \frac{6}{13} & \frac{4}{13} \end{bmatrix}\begin{bmatrix} \frac{9}{13} & \frac{6}{13} \\ \frac{6}{13} & \frac{4}{13} \end{bmatrix} = \begin{bmatrix} \frac{117}{13^2} & \frac{78}{13^2} \\ \frac{78}{13^2} & \frac{52}{13^2} \end{bmatrix} = \begin{bmatrix} \frac{9}{13} & \frac{6}{13} \\ \frac{6}{13} & \frac{4}{13} \end{bmatrix}.$$

A similar calculation verifies that $(I - AB)^2 = I - AB$ where $I - AB = \begin{bmatrix} \frac{4}{13} & -\frac{6}{13} \\ -\frac{6}{13} & \frac{9}{13} \end{bmatrix}$. To say that $BY = X$ is the best solution of $AX = Y$ means, as in the case $Y = \begin{bmatrix} 2 \\ 1 \end{bmatrix}$ above, that ABY is the point on the line through (0, 0) and

(3, 2) nearest to Y. Geometrically, ABY is the *orthogonal projection* of Y to the line. The entries of $(I - AB)Y = Y - ABY$ are the differences between the x- and y-coordinates of Y and those of its projection ABY on the line. The identity $\|ABY\|^2 + \|(I - AB)Y\|^2 = \|Y\|^2$ of Section 4 expresses the Pythagorean theorem that the square of the length of the diagonal is the sum of the squares of the lengths of the sides of the right triangle of which Y and (0, 0) are vertices and the side opposite Y lies along the given line.

The traditional application of the method of least squares is to problems in which the mathematical formula that governs a particular phenomenon is known, but the formula contains parameters that need to be determined on the basis of empirical tests. For example, suppose that x and y are two quantities that are known to be related by the formula $y = mx + b$, but the values of the parameters m and b need to be determined by performing experiments to find particular values of x and y. Say three experiments are performed and that the values $(x, y) = (0, 0.1), (1, 2.1), (2, 3.9)$ are found. What is the best choice of m and b in $y = mx + b$?

If the line fitted the observed values perfectly, the equations $m \cdot 0 + b = 0.1$, $m \cdot 1 + b = 2.1$, and $m \cdot 2 + b = 3.9$ would hold; that is, one would have $AX = Y$ where

$$A = \begin{bmatrix} 0 & 1 \\ 1 & 1 \\ 2 & 1 \end{bmatrix}, \qquad X = \begin{bmatrix} m \\ b \end{bmatrix}, \qquad Y = \begin{bmatrix} 0.1 \\ 2.1 \\ 3.9 \end{bmatrix}.$$

However, this matrix division problem has no solution—presumably because the observed values of (x, y) were slightly inaccurate—and the problem is to use all three observations to arrive at a choice of the "best" values for m and b.

Since A has rank 2 (the determinant of the first two rows is $0 \cdot 1 - 1 \cdot 1 \neq 0$), the mate of A is $B = (A^T A)^{-1} A^T$, and the method of least squares chooses the values of m and b given by

$$\begin{bmatrix} m \\ b \end{bmatrix} = X = BY = \begin{bmatrix} 5 & 3 \\ 3 & 3 \end{bmatrix}^{-1} \begin{bmatrix} 0 & 1 & 2 \\ 1 & 1 & 1 \end{bmatrix} \begin{bmatrix} 0.1 \\ 2.1 \\ 3.9 \end{bmatrix} = \begin{bmatrix} \frac{3}{6} & -\frac{3}{6} \\ -\frac{3}{6} & \frac{5}{6} \end{bmatrix} \begin{bmatrix} 9.9 \\ 6.1 \end{bmatrix} = \begin{bmatrix} \frac{19}{10} \\ \frac{4}{30} \end{bmatrix}.$$

(The inverse of a 2×2 matrix is easily found using the formula $\begin{bmatrix} a & b \\ c & d \end{bmatrix} \begin{bmatrix} d & -b \\ -c & a \end{bmatrix} = \begin{bmatrix} ad - bc & 0 \\ 0 & ad - bc \end{bmatrix}$.) Note that this answer is plausible; a slope of about 2 and a y-intercept near 0.1 give a line near the given points.

The "least squares" criterion by which this line $y = mx + b$ was chosen is that the sum of the squares of the differences of the three given y values and the three y values predicted by the formula $y = mx + b$ is to be as small as possible.

In actual applications of the method, one normally makes a large number of observations in the hope that the observational errors will cancel each other out and the method will produce very accurate values of m and b. Note that even if there are many more than three observations, the evaluation of m and b involves only additions, multiplications, and the inversion of a 2×2 matrix. More generally, if the method is being used to determine the values of k parameters, only the inversion of a $k \times k$ matrix is needed in addition to simple arithmetic, no matter how many observations are used.

The mate of $A = \begin{bmatrix} -3 & 1 & 1 \\ 1 & -1 & 1 \\ 1 & 1 & -3 \end{bmatrix}$ is $B = \frac{1}{72}\begin{bmatrix} -17 & 8 & 1 \\ 8 & -8 & 8 \\ 1 & 8 & -17 \end{bmatrix}$, as can be found in the following way. By the algorithm of Chapter 2,

$$\begin{bmatrix} 1 & 0 & 0 \\ -1 & -7 & -4 \\ 1 & 2 & 1 \end{bmatrix}\begin{bmatrix} -3 & 1 & 1 \\ 1 & -1 & 1 \\ 1 & 1 & -3 \end{bmatrix}\begin{bmatrix} 1 & -1 & 1 \\ 4 & -3 & 2 \\ 0 & 0 & 1 \end{bmatrix} = \begin{bmatrix} 1 & 0 & 0 \\ 0 & 2 & 0 \\ 0 & 0 & 0 \end{bmatrix}.$$

Computation of the inverses of the unimodular matrices in this equation gives

$$M^{-1} = \begin{bmatrix} 1 & 0 & 0 \\ -3 & 1 & 4 \\ 5 & -2 & -7 \end{bmatrix} \quad \text{and} \quad N^{-1} = \begin{bmatrix} -3 & 1 & 1 \\ -4 & 1 & 2 \\ 0 & 0 & 1 \end{bmatrix},$$

from which

$$\begin{bmatrix} -3 & 1 & 1 \\ 1 & -1 & 1 \\ 1 & 1 & -3 \end{bmatrix} = \begin{bmatrix} 1 & 0 & 0 \\ -3 & 1 & 4 \\ 5 & -2 & -7 \end{bmatrix}\begin{bmatrix} 1 & 0 & 0 \\ 0 & 2 & 0 \\ 0 & 0 & 0 \end{bmatrix}\begin{bmatrix} -3 & 1 & 1 \\ -4 & 1 & 2 \\ 0 & 0 & 1 \end{bmatrix}$$
$$= \begin{bmatrix} 1 & 0 \\ -3 & 1 \\ 5 & -2 \end{bmatrix}\begin{bmatrix} 1 & 0 \\ 0 & 2 \end{bmatrix}\begin{bmatrix} -3 & 1 & 1 \\ -4 & 1 & 2 \end{bmatrix}.$$

Let P, Q, and R be the factors in the last product. Then

$$P^M = (P^TP)^{-1}P^T = \frac{1}{6}\begin{bmatrix} 5 & 13 \\ 13 & 35 \end{bmatrix}\begin{bmatrix} 1 & -3 & 5 \\ 0 & 1 & -2 \end{bmatrix} = \frac{1}{6}\begin{bmatrix} 5 & -2 & -1 \\ 13 & -4 & -5 \end{bmatrix},$$

$$Q^M = Q^{-1} = \begin{bmatrix} 1 & 0 \\ 0 & \frac{1}{2} \end{bmatrix}, \quad \text{and}$$

$$R^M = R^T(RR^T)^{-1} = \begin{bmatrix} -3 & -4 \\ 1 & 1 \\ 1 & 2 \end{bmatrix}\frac{1}{6}\begin{bmatrix} 21 & -15 \\ -15 & 11 \end{bmatrix} = \frac{1}{6}\begin{bmatrix} -3 & 1 \\ 6 & -4 \\ -9 & 7 \end{bmatrix}.$$

Therefore, the mate of the given matrix is

$$R^MQ^MP^M = \frac{1}{6}\begin{bmatrix} -3 & 1 \\ 6 & -4 \\ -9 & 7 \end{bmatrix}\frac{1}{2}\begin{bmatrix} 2 & 0 \\ 0 & 1 \end{bmatrix}\frac{1}{6}\begin{bmatrix} 5 & -2 & -1 \\ 13 & -4 & -5 \end{bmatrix}$$
$$= \frac{1}{72}\begin{bmatrix} -17 & 8 & 1 \\ 8 & -8 & 8 \\ 1 & 8 & -17 \end{bmatrix}$$

as was stated above.

The computation can be shortened by finding the factorization

$$A = \begin{bmatrix} -3 & 1 \\ 1 & -1 \\ 1 & 1 \end{bmatrix} \begin{bmatrix} 1 & 0 & -1 \\ 0 & 1 & -2 \end{bmatrix}$$

by inspection. If the factors on the right are respectively called F and G, then $F^T F = \begin{bmatrix} 11 & -3 \\ -3 & 3 \end{bmatrix}$ and $GG^T = \begin{bmatrix} 2 & 2 \\ 2 & 5 \end{bmatrix}$, which leads to the conclusion that the mate of A is

$$\begin{bmatrix} 1 & 0 \\ 0 & 1 \\ -1 & -2 \end{bmatrix} \begin{bmatrix} \frac{5}{6} & -\frac{2}{6} \\ -\frac{2}{6} & \frac{2}{6} \end{bmatrix} \begin{bmatrix} \frac{3}{24} & \frac{3}{24} \\ \frac{3}{24} & \frac{11}{24} \end{bmatrix} \begin{bmatrix} -3 & 1 & 1 \\ 1 & -1 & 1 \end{bmatrix}.$$

Exercises

1 Find the mate of each of the following matrices.

(a) $\begin{bmatrix} 2 & 2 & 2 \end{bmatrix}$ (b) $\begin{bmatrix} 2 & 0 & 0 \\ 0 & 2 & 0 \\ 0 & 0 & 2 \end{bmatrix}$ (c) $\begin{bmatrix} 2 & 0 & 0 \\ 0 & 2 & 0 \end{bmatrix}$ (d) $\begin{bmatrix} 2 & 0 \\ 0 & 0 \end{bmatrix}$

2 Find the line $y = mx + b$ that best fits the data $(x, y) = (0, 0), (1, 0), (2, 2)$.

3 Find the "best" solution of $3x + 2y = 6$ and give a geometrical interpretation of the answer.

4 Find the "best" approximate solution (a, b) of

$$a + 3b = 1$$
$$2a - b = 3$$
$$a - b = 5.$$

5 Find the parabola $y = ax^2 + bx + c$ that best fits the data $(x, y) = (-10, 110), (0, 0), (10, 95), (20, 400)$.

6 Let $A = \begin{bmatrix} 3 \\ 2 \\ 1 \end{bmatrix}$ and $Y = \begin{bmatrix} 4 \\ 3 \\ 2 \end{bmatrix}$. Find the number t for which $\|At - Y\|^2$ is a minimum. What is the minimum value of $\|At - Y\|^2$? Describe this problem and its solution in terms of a line in three-dimensional space.

7 Let A be as in the preceding exercise, and let B be the mate of A. Let P_1 and P_2 be AB and $I - AB$, respectively. Find the mates of P_1 and P_2. Describe the geometrical significance of P_1 and P_2 and of the equations $P_i^2 = P_i$.

8 Find the mate of $\begin{bmatrix} 0 & 1 \\ 0 & 0 \end{bmatrix}$.

9 Find the mate of $A = \begin{bmatrix} 1 & -1 & 4 \\ 2 & -4 & 7 \\ 1 & 1 & 5 \end{bmatrix}$.

10 If S is a symmetric matrix and R is its mate, show that $RS = SR$.

11 The characterization of the mate of A as the unique matrix B for which $ABA = A$, $BAB = B$, and AB and BA are symmetric was given by R. Penrose in 1955. Already in 1920 E. H. Moore associated to a matrix A the unique solution B of $ABA = A$ that is divisible on both left and right by A^T. Prove that for any given matrix A the mate B of A is divisible on both the left and the right by A^T. Prove that any matrix C for which $ACA = A$ and for which there exist matrices U and V with $C = A^T U$ and $C = VA^T$ must be B.

Chapter 8
Matrices with Polynomial Entries

1 Polynomials

A **polynomial in one variable** is an expression of the form $a_n x^n + a_{n-1}x^{n-1} + \cdots + a_0$, where x is a variable*, where n is a nonnegative integer, and where a_n, $a_{n-1}, \ldots, a_0$ are numbers. (Here, and in the remainder of the book, "number" will mean "rational number" unless otherwise stated.) The a's are called the **coefficients** of the polynomial. The **leading term** of a polynomial is the term $a_i x^i$ with $a_i \neq 0$ for which i is as large as possible. Normally one assumes $a_n \neq 0$, so $a_n x^n$ is the leading term. The **leading coefficient** of a polynomial is the coefficient of the leading term; the **degree** is the exponent of x in the leading term. A polynomial of degree 0 is simply a nonzero rational number. The polynomial 0 has no leading term; it is considered to have degree $-\infty$ in order to make the degree of the product of two polynomials equal to the sum of their degrees in all cases. For a similar reason, the leading coefficient of the polynomial 0 is considered to be 0. A polynomial is called **monic** if its leading coefficient is 1.

Polynomials are added and multiplied in the familiar ways. Briefly put, the sum of two polynomials is found by using the commutative law of addition to rearrange terms and using the distributive law $ax^i + bx^i = (a+b)x^i$ to consolidate terms in which x occurs to the same power. The product of two polynomials is found by using the distributive law and the rule $(ax^i)(bx^j) = abx^{i+j}$ to write $(a_0x^n + a_1x^{n-1} + \cdots + a_n)(b_0x^m + b_1x^{m-1} + \cdots + b_m)$ as a sum of $(m+1)(n+1)$ terms, which can then be added. A polynomial $f(x)$ is a **multiple** of another $g(x)$ if there is a third polynomial $h(x)$ such that $f(x) = g(x)h(x)$.

The reader probably has some experience in computation with polynomials, including, perhaps, methods of organizing the operation of multiplication of polynomials and an algorithm for long division of polynomials. Only very simple computations with polynomials will be needed.

---

*More properly, x is an *indeterminate,* a letter one calculates with, rather than a variable quantity that will later be given specific values. However, the word "variable" is customarily used.

2 Equivalence of Matrices of Polynomials

A matrix with polynomial entries is simply an $m \times n$ array of polynomials. Such matrices are multiplied (when the number of columns of the left factor is equal to the number of rows in the right factor) in exactly the same way that matrices with integer entries or matrices with rational entries are. The meaning of this operation is, of course, connected with the interpretation of the matrices as matrices of coefficients of linear substitutions, as in Chapter 1. However, because the variable x in the entries of the matrix could be confused with the variables of the substitutions, we will deal with them as we have with all matrices since Chapter 1 purely as matrices.

We will say that two matrices with polynomial entries are **equivalent** if one can be transformed into the other by a sequence of operations in which a multiple of a row is added to an adjacent row or a multiple of a column is added to an adjacent column.

At first glance, this definition is very different from the definition of equivalence that has been used so far in the book; until now, the basic operations have been addition or subtraction of the row (column) itself, whereas now it is addition of a *multiple* of the row (column). Since subtraction of a row or column is the same as addition of -1 times that row or column, the operations that are now allowed include the operations that were used before. Conversely, addition of any *integer* multiple of a row is the result of repeated additions (if the integer is positive) or repeated subtractions (if the integer is negative) of the row, and the same for columns. Therefore, for matrices with *integer* entries each operation of one type can be achieved using operations of the other type, and the definition of Chapter 2 could also have been stated in exactly the same way as the definition just given for matrices with polynomial entries. The definition used in Chapter 2 was chosen because addition and subtraction are more basic operations than multiplication. For matrices with *rational* entries—if "multiple" is taken in the above definition to mean rational multiples—the two definitions are different, as will be seen in Section 8. (See also Exercise 7 of Chapter 6.) For matrices with *polynomial* entries, the two definitions are quite different; the new definition in terms of the addition of multiples is the only one of interest, because the old definition does not have the property that every matrix is equivalent to a diagonal one.

3 Unimodular Matrices

Let A be an $m \times n$ matrix whose entries are polynomials in one variable. The operation of adding $f(x)$ times the first row of A to the second row is the same as the operation of multiplying A on the left by the matrix obtained by adding $f(x)$ times the first row of I_m to the second, as follows immediately from the definition of matrix product. Similarly, the operation of addition of $f(x)$ times any row of A to an adjacent row can be achieved by applying the desired operation to I_m and multiplying A on the left by the result; in the same way, the operation of addition of $f(x)$ times any column of A to an adjacent column can be achieved by applying the desired operation to I_n and multiplying A on the right by the result.

We will call a matrix a **polynomial tilt** if it can be obtained from an identity matrix by adding a polynomial times one row to an adjacent row, or, what is the same, if it can be obtained from an identity matrix by adding a polynomial times one column to an adjacent column. We will call a matrix of polynomials **unimodular** if it is equivalent to the identity matrix, or, what is the same, if it is a product of polynomial tilts.* A polynomial tilt has an obvious inverse (reverse the sign of the single nonzero entry off the diagonal), so a unimodular matrix is invertible. Saying that two matrices of polynomials A and B are equivalent is the same as saying that there exist unimodular matrices M and N such that $B = MAN$.

4 The 1 × 2 Case

Theorem. *A 1×2 matrix whose entries are polynomials in one variable is equivalent to a matrix in which the entry on the right is zero.*

Algorithm: As long as the entry of the matrix on the right is not zero, perform the operation determined by:

1. If the entry on the left is zero, add the entry on the right to the entry on the left.

2. If the entries are of equal degree and the entry on the left is monic, use the entry on the left to reduce the degree of the entry on the right. Specifically, add $-b$ times the left entry to the right entry, where b is the leading coefficient of the right entry.

*As in Chapter 2, an identity matrix, including I_1, is to be regarded as a product of *no* polynomial tilts.

3. If the entries are of equal degree and the entry on the left is not monic, use the entry on the right to make the entry on the left monic. Specifically, add $(1-a)/b$ times the entry on the right to the entry on the left, where a is the leading coefficient of the left entry and b the leading coefficient of the right entry.

4. Otherwise, use the entry of lower degree to reduce the degree of the other entry without reducing the other entry to zero. Specifically, add $-bx^j/a$ times the entry of lower degree to the other entry, where a is the leading coefficient of the entry of lower degree, b is the leading coefficient of the entry of higher degree, and j is the difference of the degrees, unless the result of this operation would be zero, in which case add $1 - (bx^j/a)$ times the entry of lower degree to the other entry.

The algorithm could be simplified somewhat if only the 1×2 case mattered, but the proof of the $m \times n$ case in the next section and the reduction to strongly diagonal form in Section 6 are simpler when the algorithm just given is used.

Proof that the algorithm terminates. A step of type 1 is never called for, except at the first step, because the other steps never reduce the left entry to zero. A step of type 3 is followed by a step of type 2. A step of type 2 or 4 reduces the degree of one of the polyonomials. Therefore, only a finite number of steps can be executed before the algorithm terminates with the right entry equal to zero.

5 *The General Case*

Theorem. *Every matrix whose entries are polynomials in one variable is equivalent to a diagonal matrix.*

Proof. The algorithm for the 1×2 case can be used to reduce all nondiagonal entries of the matrix to zero using the same priorities as were used in Chapter 2. The proof that this algorithm terminates is essentially the same as in Chapter 2. Because the 1×2 algorithm terminates, the general algorithm must reach a matrix in which all entries of the first row after the first are zero. The subsequent operations (if any) required by nonzero entries in the first column either reduce all entries of the first column after the first to zero or change the first entry of the first column. The first entry of the first column can change only from zero to a nonzero polynomial, or from a nonmonic polynomial to a monic polynomial, or from a nonzero polynomial to a nonzero polynomial of *lower* degree. Therefore, the first entry can only change a finite number of times, and all entries in the first row and column not on the diagonal must eventually become zero. The subsequent operations of the algorithm change only a matrix with one less row and one less column, so, for the same reason, they must eventually reduce all nondiagonal entries in the first row and column of the smaller matrix to zero, and so forth, until all nondiagonal entries are reduced to zero.

6 *Determinants*

A **determinant** of a square matrix whose entries are polynomials in one variable is the product of the diagonal entries of an equivalent diagonal matrix. As in the case of matrices with integer entries, *a square matrix has only one determinant* because one can give a *formula* for the determinant of an $n \times n$ matrix. The formula is in fact the same formula as in Chapter 4. The determinant is equal to the sum of $n!$ terms, each of which is plus or minus a product of n entries of the matrix with one entry from each row and one entry from each column; specifically, the sign to be associated with each of the $n!$ ways of choosing n entries with one from each row and one from each column is the sign of the determinant of the matrix which has ones in the n chosen locations and zeros everywhere else.

As in Chapter 4, the determinant of a matrix can be evaluated by finding an equivalent matrix in which one row or column has only one nonzero entry and using the expression of such a determinant in terms of the determinant of an $(n-1) \times (n-1)$ matrix, then expressing this in terms of the determinant of an $(n-2) \times (n-2)$ matrix, and so forth.

Also as in Chapter 4, the determinant of the product of two square matrices is the product of their determinants. This can be proved by proving it first for the case in which one of the matrices is diagonal and the case in which one of them is a polynomial tilt.

7 *Strongly Diagonal Form*

We will say that a matrix whose entries are polynomials in one variable is **strongly diagonal** if it is diagonal and (1) each diagonal entry after the first is a multiple of its predecessor on the diagonal and (2) each nonzero diagonal entry is monic, except that there is no such restriction on the last diagonal entry of a square matrix.

Theorem. *Every matrix whose entries are polynomials in one variable is equivalent to a strongly diagonal matrix.*

Proof. As in Chapter 5, this will be proved by augmenting the algorithm to include further operations to be performed on diagonal matrices that are not strongly diagonal.

5. If there is a diagonal entry that is not a multiple of its predecessor on the diagonal, add row i to row $i-1$, where the ith diagonal entry is the first one that fails to be a multiple of its predecessor.

6. If 5 does not determine an operation and if there is a diagonal entry not in the last column that is nonzero and not monic, add column i to column $i+1$, where the ith diagonal entry is the first one that fails to be monic.

7. If neither 5 nor 6 determines an operation and if there is a diagonal entry not in the last row that is nonzero and not monic, add row i to row $i+1$, where the ith diagonal entry is the first one that fails to be monic.

The proof that this algorithm terminates is a simple modification of the proof of the analogous fact in Chapter 5. If rule 5 applies to a matrix in which the $(i-1)$st diagonal entry is zero, then the affected part of the matrix goes through the changes

$$\begin{bmatrix} 0 & 0 \\ 0 & f_i \end{bmatrix} \sim \begin{bmatrix} 0 & f_i \\ 0 & f_i \end{bmatrix} \sim \begin{bmatrix} f_i & f_i \\ f_i & f_i \end{bmatrix} \sim \begin{bmatrix} f_i & 0 \\ f_i & 0 \end{bmatrix} \sim \begin{bmatrix} f_i & 0 \\ 0 & 0 \end{bmatrix}$$

if f_i is monic. If the leading coefficient of f_i is $c \neq 1$, the steps after the second are

$$\sim \begin{bmatrix} f_i - \frac{(c-1)f_i}{c} & f_i \\ f_i - \frac{(c-1)f_i}{c} & f_i \end{bmatrix} = \begin{bmatrix} \frac{f_i}{c} & f_i \\ \frac{f_i}{c} & f_i \end{bmatrix} \sim \begin{bmatrix} \frac{f_i}{c} & 0 \\ \frac{f_i}{c} & 0 \end{bmatrix} \sim \begin{bmatrix} \frac{f_i}{c} & 0 \\ 0 & 0 \end{bmatrix}.$$

If rule 5 applies to a matrix in which the $(i-1)$st diagonal entry is not zero, the affected part of the matrix goes through the changes

$$\begin{bmatrix} f_{i-1} & 0 \\ 0 & f_i \end{bmatrix} \sim \begin{bmatrix} f_{i-1} & f_i \\ 0 & f_i \end{bmatrix} \sim \cdots \sim \begin{bmatrix} r & 0 \\ s & t \end{bmatrix} \sim \cdots \sim \begin{bmatrix} u & 0 \\ 0 & v \end{bmatrix}$$

where $\begin{bmatrix} r & 0 \\ s & t \end{bmatrix}$ represents the first matrix in the sequence with a zero in the upper right corner. The operations that precede the first appearance of a zero in the upper right corner are all column operations, so all entries in the first row of all the matrices up to this point are multiples of r. In particular, f_{i-1} and f_i are multiples of r, which implies $r \neq 0$. Moreover, r is monic because it must have been used to reduce the upper right corner to zero. If r did not have lower degree than f_{i-1}, then f_{i-1} would be a nonzero rational number times r, which, because f_i is a multiple of r, would be contrary to the assumption that f_i is not a multiple of f_{i-1}. Therefore, r must have lower degree than f_{i-1}. Since both f_{i-1} and f_i are multiples of r, all entries of all matrices of the sequence are multiples of r. In particular, s is a multiple of r. The subsequent operations of the algorithm therefore merely reduce s to zero without changing r or t, that is, $r = u$ and $t = v$. Thus, the transition reduces the degree of the first diagonal entry it changes. However, it leaves the product of the two affected diagonal entries unchanged—that is, $f_{i-1}f_i = uv$—because equivalent matrices have equal determinants.

Thus, as in Chapter 5, if one considers just the diagonal matrices in the sequence of equivalent matrices generated by the algorithm, the transition from one diagonal matrix to the next either moves the first zero entry on the diagonal to the next diagonal position without changing the degree of the diagonal entry with which it is interchanged, or reduces the sum of the degrees of the first i nonzero diagonal entries for some i without increasing this sum for any other i. (The sum of the degrees of a set of polynomials is to be regarded here as being analogous to the absolute value of the product of a set of integers.) Since there are at most a finite number of changes of this type, the algorithm must eventually arrive at a diagonal matrix in which each diagonal entry is a multiple of its predecessor on the diagonal, and only rule 6 or rule 7 can be invoked.

When rule 6 is invoked, the transition to the next diagonal matrix is, in the 2×2 matrix where all the changes take place,

$$\begin{bmatrix} f_i & 0 \\ 0 & f_{i+1} \end{bmatrix} \sim \begin{bmatrix} f_i & f_i \\ 0 & f_{i+1} \end{bmatrix} \sim \begin{bmatrix} f_i + \frac{(1-c)f_i}{c} & f_i \\ \frac{(1-c)f_{i+1}}{c} & f_{i+1} \end{bmatrix}$$
$$\sim \begin{bmatrix} \frac{f_i}{c} & 0 \\ \frac{(1-c)f_{i+1}}{c} & f_{i+1} - c\frac{(1-c)f_{i+1}}{c} \end{bmatrix} \sim \begin{bmatrix} \frac{f_i}{c} & 0 \\ 0 & cf_{i+1} \end{bmatrix}.$$

In short, the first nonzero diagonal entry that fails to be monic is divided by its leading coefficient c, the following diagonal entry is multiplied by c, and nothing else is changed. The ultimate effect of an operation resulting from rule 7 is even simpler—a nonmonic diagonal entry is merely divided by its leading coefficient—so these operations make the first nonmonic diagonal entry monic without changing the fact that each diagonal entry after the first is a multiple of its predecessor. After a finite number of such changes, the matrix becomes strongly diagonal and the algorithm terminates.

Theorem. *Two strongly diagonal matrices are equivalent only if they are equal.*

Proof. As in Chapter 5, the essential observation is that if A is equivalent to the strongly diagonal matrix with diagonal entries $a_1(x), a_2(x), \ldots, a_r(x)$, then all $k \times k$ minors of A are multiples of $a_1(x)a_2(x)\cdots a_k(x)$ (Exercise 6). Therefore, if two strongly diagonal matrices are equivalent, they must have the same number of nonzero diagonal entries. If two *monic* polynomials have the property that each is a multiple of the other, then they are equal. Therefore, the product of the first k nonzero diagonal entries of two equivalent strongly diagonal matrices are the same for all k. (Except in the case $k = m = n$, this follows from the fact that these nonzero diagonal entries are all monic; in the remaining case it follows from the fact that equivalent matrices have equal determinants.) By division, then, all nonzero diagonal entries are equal, and the theorem follows.

Corollary. *Given a set of polynomials in x, say* $f_1(x), f_2(x), \ldots, f_n(x)$, *not all of which are zero, there is a unique monic polynomial* $g(x)$ *with the properties (1) each* $f_i(x)$ *is a multiple of* $g(x)$ *and (2)* $g(x)$ *is a sum of multiples of* $f_1(x), f_2(x), \ldots, f_n(x)$.

Proof. If $n = 1$, $g(x)$ is the polynomial obtained by dividing $f_1(x)$ by its leading coefficient. If $n > 1$, let $h(x)$ be the first entry of the unique strongly diagonal matrix equivalent to $A = [\, f_1(x) \quad f_2(x) \quad \cdots \quad f_n(x)\,]$. It follows from $A \neq 0$ that $h(x) \neq 0$ and therefore that $h(x)$ is monic. Since $h(x)$ is the first entry of AN where N is a unimodular matrix, $h(x)$ is a sum of multiples of the entries of A. (The multipliers are the entries of the first column of N.) Since each $f_i(x)$ is an entry of $[\, h(x) \quad 0 \quad \cdots \quad 0\,]\, N^{-1}$, each $f_i(x)$ is a multiple of $h(x)$. (The multiplier is the corresponding entry of the first row of N^{-1}.) Thus, $h(x)$ is a monic polynomial with the two required properties. If $g(x)$ is any monic polynomial with these two properties, then, because $g(x)$ is a sum of multiples of the entries of A and each entry of A is a multiple of $h(x)$, $g(x)$ is a multiple of $h(x)$. For the same reason, $h(x)$ is a multiple of $g(x)$. Since both are monic, it follows that $g(x) = h(x)$, as was to be shown.

The unique monic polynomial $g(x)$ of the corollary is called the **greatest common divisor** of the polynomials $f_1(x), f_2(x), \ldots, f_n(x)$, abbreviated gcd.

8 Wider Equivalence

We will say that two matrices whose entries are rational numbers are **equivalent in the wider sense** if they are equivalent as matrices with polynomial entries. Because the algorithm for reducing a matrix of polynomials to strongly diagonal form never introduces x if x does not appear in the matrix to start with, this algorithm reduces a matrix of rational numbers to a matrix of rational numbers which, when regarded as a matrix of polynomials, is strongly diagonal. Clearly an $m \times n$ matrix of rational numbers is strongly diagonal as a matrix of polynomials only if (1) for some k it is a matrix whose first k diagonal entries are 1 and whose remaining entries are zero, or (2) it is an $n \times n$ diagonal matrix whose first $n - 1$ diagonal entries are 1. Therefore:

Theorem. *A matrix whose entries are rational numbers is equivalent in the wider sense to a matrix of one of the two types just described. Two such matrices are equivalent in the wider sense only if they are equal.*

The difference between ordinary equivalence and equivalence in the wider sense can be seen clearly if one considers the effect of the two reduction algorithms on the matrix $[\, 2 \quad 0\,]$. The algorithm of Chapter 5 leaves this matrix unchanged, because it is already strongly diagonal. The algorithm of the present chapter changes it to $[\,2 \quad 2\,]$, then to $[\,1 \quad 2\,]$, then to $[\,1 \quad 0\,]$. The main difference is that *in the wider sense* $[\,2 \quad 2\,]$ is equivalent to $[\,1 \quad 2\,]$ (subtract half the right entry from the left entry) but in the ordinary sense these matrices of rational numbers are not equivalent (the equivalent strongly diagonal matrix is $[\,2 \quad 0\,]$ in the case of the first and $[\,1 \quad 0\,]$ in the case of the second).

We will say that a square matrix whose entries are rational numbers is **unimodular in the wider sense** if it is unimodular as a matrix whose entries are polynomials, or, what is the same, if its determinant is 1.

Examples

The algorithm of Section 4 applied to $[\,x^2+2x+1 \quad x^3+x^2+x+1\,]$ gives the sequence

$$\sim [\,x^2+2x+1 \quad -x^2+1\,] \sim [\,x^2+2x+1 \quad 2x+2\,] \sim [\,x+1 \quad 2x+2\,] \sim [\,x+1 \quad 0\,]$$

terminating with a diagonal matrix. Since $x+1$ is monic, the final matrix is in fact strongly diagonal, and $x+1$ is the gcd of x^2+2x+1 and x^3+x^2+x+1.

The matrices $\begin{bmatrix} 2 & 0 \\ 0 & 2 \end{bmatrix}$ and $\begin{bmatrix} 1 & 0 \\ 0 & 4 \end{bmatrix}$ are strongly diagonal and not equal, so they are not equivalent. On the other hand, they are equivalent in the wider sense, because adding half the second row to the first in $\begin{bmatrix} 2 & 0 \\ 0 & 2 \end{bmatrix}$ shows that it is equivalent in the wider sense to $\begin{bmatrix} 2 & 1 \\ 0 & 2 \end{bmatrix}$. Since this matrix is equivalent to $\begin{bmatrix} 1 & 0 \\ 0 & 4 \end{bmatrix}$, and since equivalence implies equivalence in the wider sense, the two given matrices are equivalent in the wider sense.

The next chapter makes use of matrices of polynomials of the special form $xI - A$, in which A is an $n \times n$ matrix of rational numbers, I is the $n \times n$ identity matrix, and x is the variable. The following examples are of this special form.

Let A be the 3×3 matrix $\begin{bmatrix} 0 & 0 & 0 \\ 1 & 0 & 0 \\ 0 & 1 & 0 \end{bmatrix}$. Then

$$xI - A = \begin{bmatrix} x & 0 & 0 \\ -1 & x & 0 \\ 0 & -1 & x \end{bmatrix} \sim \begin{bmatrix} -1 & x^2+x & 0 \\ -1 & x & 0 \\ 0 & -1 & x \end{bmatrix} \sim \begin{bmatrix} -1 & x & 0 \\ -1 & -x^2+x & 0 \\ 0 & -1 & x \end{bmatrix} \sim$$

$$\sim \begin{bmatrix} -1 & -1 & 0 \\ -1 & -x^2-1 & 0 \\ 0 & -1 & x \end{bmatrix} \sim \begin{bmatrix} 1 & -1 & 0 \\ 2x^2+1 & -x^2-1 & 0 \\ 2 & -1 & x \end{bmatrix} \sim \begin{bmatrix} 1 & 0 & 0 \\ 2x^2+1 & x^2 & 0 \\ 2 & 1 & x \end{bmatrix} \sim$$

$$\sim \begin{bmatrix} 1 & 0 & 0 \\ 1 & 0 & -x^3 \\ 2 & 1 & x \end{bmatrix} \sim \begin{bmatrix} 1 & 0 & 0 \\ 1 & 0 & -x^3 \\ 0 & 1 & 2x^3+x \end{bmatrix} \sim \begin{bmatrix} 1 & 0 & 0 \\ 0 & 0 & -x^3 \\ 0 & 1 & 2x^3+x \end{bmatrix} \sim$$

$$\sim \begin{bmatrix} 1 & 0 & 0 \\ 0 & -x^3 & -x^3 \\ 0 & 2x^3+x+1 & 2x^3+x \end{bmatrix} \sim \begin{bmatrix} 1 & 0 & 0 \\ 0 & x^3 & -x^3 \\ 0 & -2x^3-x+1 & 2x^3+x \end{bmatrix} \sim$$

$$\sim \begin{bmatrix} 1 & 0 & 0 \\ 0 & x^3 & 0 \\ 0 & -2x^3-x+1 & 1 \end{bmatrix} \sim \begin{bmatrix} 1 & 0 & 0 \\ 0 & x^3 & 0 \\ 0 & -x+1 & 1 \end{bmatrix} \sim$$

$$\sim \begin{bmatrix} 1 & 0 & 0 \\ 0 & x^2 & x^2 \\ 0 & -x+1 & 1 \end{bmatrix} \sim \begin{bmatrix} 1 & 0 & 0 \\ 0 & x^2 & 0 \\ 0 & -x+1 & x \end{bmatrix} \sim \begin{bmatrix} 1 & 0 & 0 \\ 0 & x & x^2 \\ 0 & -x+1 & x \end{bmatrix} \sim$$

$$\sim \begin{bmatrix} 1 & 0 & 0 \\ 0 & x & 0 \\ 0 & -x+1 & x^2 \end{bmatrix} \sim \begin{bmatrix} 1 & 0 & 0 \\ 0 & x & 0 \\ 0 & 1 & x^2 \end{bmatrix} \sim \begin{bmatrix} 1 & 0 & 0 \\ 0 & 1 & -x^3+x^2 \\ 0 & 1 & x^2 \end{bmatrix} \sim$$

$$\sim \begin{bmatrix} 1 & 0 & 0 \\ 0 & 1 & x^2 \\ 0 & 1 & x^3+x^2 \end{bmatrix} \sim \begin{bmatrix} 1 & 0 & 0 \\ 0 & 1 & 0 \\ 0 & 1 & x^3 \end{bmatrix} \sim \begin{bmatrix} 1 & 0 & 0 \\ 0 & 1 & 0 \\ 0 & 0 & x^3 \end{bmatrix}.$$

The strongly diagonal matrix equivalent to $xI - A$ in this case can be found much more easily by noting that one of its 2×2 minors is $\begin{vmatrix} -1 & x \\ 0 & -1 \end{vmatrix} = 1$. Therefore the strongly diagonal matrix equivalent to it must have 1 as its first two diagonal entries; the last diagonal entry must therefore be $|xI - A| = x^3$, as was found above.

Let A be a 3×3 matrix of zeros except that the middle entry in the last row is 1. Then

$$xI - A = \begin{bmatrix} x & 0 & 0 \\ 0 & x & 0 \\ 0 & -1 & x \end{bmatrix} \sim \begin{bmatrix} x & 0 & 0 \\ 0 & -1 & x^2+x \\ 0 & -1 & x \end{bmatrix} \sim \begin{bmatrix} x & 0 & 0 \\ 0 & -1 & x \\ 0 & -1 & -x^2+x \end{bmatrix} \sim$$

$$\sim \begin{bmatrix} x & 0 & 0 \\ 0 & -1 & -1 \\ 0 & -1 & -x^2-1 \end{bmatrix} \sim \begin{bmatrix} x & 0 & 0 \\ 0 & 1 & -1 \\ 0 & 2x^2+1 & -x^2-1 \end{bmatrix} \sim \begin{bmatrix} x & 0 & 0 \\ 0 & 1 & 0 \\ 0 & 2x^2+1 & x^2 \end{bmatrix} \sim$$

$$\sim \begin{bmatrix} x & 0 & 0 \\ 0 & 1 & 0 \\ 0 & 1 & x^2 \end{bmatrix} \sim \begin{bmatrix} x & 0 & 0 \\ 0 & 1 & 0 \\ 0 & 0 & x^2 \end{bmatrix} \sim \begin{bmatrix} x & 1 & 0 \\ 0 & 1 & 0 \\ 0 & 0 & x^2 \end{bmatrix} \sim \begin{bmatrix} 1 & 1 & 0 \\ -x+1 & 1 & 0 \\ 0 & 0 & x^2 \end{bmatrix} \sim$$

$$\sim \begin{bmatrix} 1 & 0 & 0 \\ -x+1 & x & 0 \\ 0 & 0 & x^2 \end{bmatrix} \sim \begin{bmatrix} 1 & 0 & 0 \\ 1 & x & 0 \\ 0 & 0 & x^2 \end{bmatrix} \sim \begin{bmatrix} 1 & 0 & 0 \\ 0 & x & 0 \\ 0 & 0 & x^2 \end{bmatrix}.$$

Again, the strongly diagonal matrix equivalent to this $xI - A$ can be found more easily by noting that the gcd of the 2×2 minors of $xI - A$ is x and that $|xI - A|$ is x^3.

When A is the 3×3 matrix of zeros, $xI - A = \begin{bmatrix} x & 0 & 0 \\ 0 & x & 0 \\ 0 & 0 & x \end{bmatrix}$ is itself strongly diagonal. The observation that the strongly diagonal matrices in these three examples all have determinant x^3 and that they are the only strongly diagonal matrices with this determinant will be important in Chapter 9.

Let A be the 3×3 matrix $\begin{bmatrix} 3 & 1 & 1 \\ 1 & 5 & 1 \\ 1 & 1 & 3 \end{bmatrix}$. Then

$$xI - A = \begin{bmatrix} x-3 & -1 & -1 \\ -1 & x-5 & -1 \\ -1 & -1 & x-3 \end{bmatrix} \sim \begin{bmatrix} x-3 & 1 & -1 \\ -1 & x-3 & -1 \\ -1 & -2x+5 & x-3 \end{bmatrix} \sim$$

$$\sim \begin{bmatrix} x-3 & 1 & 0 \\ -1 & x-3 & x-4 \\ -1 & -2x+5 & -x+2 \end{bmatrix} \sim \begin{bmatrix} -3 & 1 & 0 \\ -x^2+3x-1 & x-3 & x-4 \\ 2x^2-5x-1 & -2x+5 & -x+2 \end{bmatrix} \sim$$

$$\sim \begin{bmatrix} 1 & 1 & 0 \\ -x^2+7x-13 & x-3 & x-4 \\ 2x^2-13x+19 & -2x+5 & -x+2 \end{bmatrix} \sim$$

$$\sim \begin{bmatrix} 1 & 0 & 0 \\ -x^2+7x-13 & x^2-6x+10 & x-4 \\ 2x^2-13x+19 & -2x^2+11x-14 & -x+2 \end{bmatrix} \sim$$

$$\sim \begin{bmatrix} 1 & 0 & 0 \\ 0 & x^2-6x+10 & x-4 \\ 0 & -2x^2+11x-14 & -x+2 \end{bmatrix} \sim \begin{bmatrix} 1 & 0 & 0 \\ 0 & -2x+10 & x-4 \\ 0 & -x^2+9x-14 & -x+2 \end{bmatrix} \sim$$

$$\sim \begin{bmatrix} 1 & 0 & 0 \\ 0 & x-2 & x-4 \\ 0 & -x^2+6x-8 & -x+2 \end{bmatrix} \sim \begin{bmatrix} 1 & 0 & 0 \\ 0 & x-2 & -2 \\ 0 & -x^2+6x-8 & x^2-7x+10 \end{bmatrix} \sim$$

$$\sim \begin{bmatrix} 1 & 0 & 0 \\ 0 & -2 & -2 \\ 0 & \frac{1}{2}x^3-\frac{9}{2}x^2+11x-8 & x^2-7x+10 \end{bmatrix} \sim$$

$$\sim \begin{bmatrix} 1 & 0 & 0 \\ 0 & 1 & -2 \\ 0 & \frac{1}{2}x^3-6x^2+\frac{43}{2}x-23 & x^2-7x+10 \end{bmatrix} \sim$$

$$\sim \begin{bmatrix} 1 & 0 & 0 \\ 0 & 1 & 0 \\ 0 & \frac{1}{2}x^3-6x^2+\frac{43}{2}x-23 & x^3-11x^2+36x-36 \end{bmatrix} \sim$$

$$\sim \begin{bmatrix} 1 & 0 & 0 \\ 0 & 1 & 0 \\ 0 & 0 & x^3-11x^2+36x-36 \end{bmatrix}.$$

(The sixth step departs from the algorithm and subtracts multiples of the first row from the second and third rows to reduce the first entries of these rows to zero. Subtraction of a multiple of the first row from the third is justified by the observation that this operation can be achieved by multiplication on the left by a unimodular matrix—see Exercise 8.)

This computation, too, can be avoided by observing that both

$$\begin{vmatrix} x-3 & -1 \\ -1 & x-5 \end{vmatrix} = \begin{vmatrix} 0 & -1 \\ (x-3)(x-5)-1 & x-3 \end{vmatrix} = (x-3)(x-5)-1 = x^2-8x+14$$

and $\begin{vmatrix} -1 & x-5 \\ -1 & -1 \end{vmatrix} = \begin{vmatrix} -1 & x-5 \\ 0 & -x+4 \end{vmatrix} = x-4$ are 2×2 minors of the matrix; therefore, the gcd of the 2×2 minors is a monic polynomial that divides $x-4$ and $x^2-8x+14-x(x-4) = -4x+14$, and $4(x-4)+(-4x+14) = -2$ and therefore must be 1. Then the equivalent strongly diagonal matrix has 1 as its first two diagonal entries and the determinant as its last entry. Thus, the entire calculation is reduced to the evaluation of $|xI - A|$, which is easy (Exercise 7).

Exercises

1 The algorithm of the text easily proves that $\begin{bmatrix} x^2+x+1 & x^2-x+1 \end{bmatrix} \sim \begin{bmatrix} 1 & 0 \end{bmatrix}$. Use the method of Chapter 2 (Section 8) to find a 2×2 matrix of polynomials N such that

$$\begin{bmatrix} x^2+x+1 & x^2-x+1 \end{bmatrix} N = \begin{bmatrix} 1 & 0 \end{bmatrix}.$$

2 Find the greatest common divisor of $8x^3+1$ and $4x^2-1$, and write it as a sum of multiples of these two polynomials.

3 Find a strongly diagonal matrix of polynomials equivalent to $\begin{bmatrix} x^2-1 & x+1 \\ 0 & 2x+2 \end{bmatrix}$.

4 The reduction of $\begin{bmatrix} x-3 & -1 & -1 \\ -1 & x-5 & -1 \\ -1 & -1 & x-3 \end{bmatrix}$ to strongly diagonal form in the examples section does not follow the algorithm. Do the computation following the algorithm. (A long calculation.)

5 Give a 2×2 matrix of rational numbers that is unimodular in the wider sense but not unimodular.

6 Prove that if D is strongly diagonal and $A \sim D$, then every $k \times k$ minor of A is divisible by the product of the first k diagonal entries of D.

7 Evaluate $\begin{vmatrix} x-3 & -1 & -1 \\ -1 & x-5 & -1 \\ -1 & -1 & x-3 \end{vmatrix}$.

8 Prove that adding a multiple of a row (column) of a matrix of polynomials to any other row (column) gives an equivalent matrix.

Chapter 9
Similarity of Matrices

1 Introduction

IN APPLICATIONS of linear algebra, the linear substitutions that arise often describe transformations of an n-dimensional space. A "transformation of an n-dimensional space" is a function that changes a quantity described by n numbers (mathematicians habitually think of such a quantity as a "point" in "n-dimensional space" even when $n > 3$) into a new quantity of the same kind. A linear substitution with n new variables and n original variables can be thought of as describing such a transformation by giving the relation between the original values of the n numbers and their values after the transformation is applied. The matrix of coefficients of a linear substitution that arises in this way is *square.* This is one reason that square matrices are especially important.

When a square matrix describes a transformation in this way, it does so only in relation to a choice of coordinates on the n-dimensional space, that is, in relation to a particular way of using n numbers to describe the quantities in question. One of the most fundamental and fruitful techniques of mathematics is to *choose new coordinates* for the description of a given problem in order to simplify it. This idea underlies the notion of **similarity of matrices**. Briefly put, two square matrices are similar if they describe the *same* transformation using *different coordinates.* To find a matrix of a simple form that is similar to a given matrix—which is the main problem studied in this chapter—amounts, therefore, to finding new coordinates that simplify the expression of the transformation the matrix represents.

The precise definition of similarity of matrices is: Let A and B be square matrices of numbers. To say that A and B are **similar**—which we will denote $A \simeq B$—means that there is an invertible square matrix of numbers P such that $B = P^{-1}AP$.

This definition relates to the informal notion of a "change of coordinates" in the following way. Let the $n \times n$ matrix A be regarded as describing a transformation of n-dimensional space relative to a given set of coordinates, and let the invertible $n \times n$ matrix P be regarded as describing a change of coordinates in the space. Then $P^{-1}AP$ describes the *same* transformation referred to the *new* coordinates insofar as it represents a composition of P^{-1} (change from the new coordinates back to the old ones) followed by A (the description of the transformation relative to the old coordinates) followed by P (to find the new coordinates of the resulting point).

The notions of transformations of n-dimensional spaces and changes of coordinates are introduced here only to give some motivation for the definition of similarity of square matrices and some indication of the reason that similarity of square matrices arises so often and is so useful in applications of linear algebra. There is no further mention of n-dimensional space or coordinate changes in the chapter.

Note that this discussion concerns matrices of numbers, not matrices of polynomials. Much use is made in this chapter of matrices of polynomials, but only as tools in the study of similarity of matrices of numbers. Specifically, as is shown in the next two sections, two square matrices of numbers satisfy $A \simeq B$ if and only if $xI - A \sim xI - B$.

2 A Necessary Condition for Similarity

Theorem. *If A and B are similar $n \times n$ matrices of rational numbers, then $xI_n - A$ and $xI_n - B$ are equivalent matrices of polynomials.*

Proof. It will first be shown that if E is any $n \times n$ matrix of rational numbers, and if D is an invertible diagonal matrix of polynomials, then $D^{-1}ED \sim E$. Because E is equivalent to a diagonal matrix, $E = T_1T_2 \cdots T_kFT_{k+1} \cdots T_l$, where each T_i is a polynomial tilt and F is diagonal. Now

$$\begin{aligned} D^{-1}ED &= D^{-1}T_1T_2 \cdots T_kFT_{k+1} \cdots T_lD^{-1} \\ &= D^{-1}T_1DD^{-1}T_2D \cdots D^{-1}T_kDD^{-1}FDD^{-1}T_{k+1}D \cdots D^{-1}T_lD. \end{aligned}$$

The right-hand side of this equation is a product of matrices of the form $D^{-1}GD$. The entry in the ith row of the jth column of $D^{-1}GD$ is $d_i^{-1}d_j$ times the entry in the ith row of the jth column of G, where d_i is the ith diagonal entry of the diagonal matrix D. This observation shows that $D^{-1}FD = F$ and that $D^{-1}T_iD$ is a polynomial tilt. Therefore, the formula just given shows—because multiplication on the left or right by a polynomial tilt $D^{-1}T_iD$ gives an equivalent matrix—that $D^{-1}ED \sim D^{-1}FD = F$. Since $E \sim F$, $D^{-1}ED \sim E$ follows.

Suppose now that $A \simeq B$, say $A = P^{-1}BP$ where P is an invertible $n \times n$ matrix of rational numbers. Because P is equivalent, as a matrix of rational numbers, to a diagonal matrix, there are unimodular matrices of integers M and N such that MPN is an invertible diagonal matrix whose entries are rational numbers, say $MPN = D$. Since M is a product of ordinary tilts, it is also a product of polynomial tilts, which implies $ME \sim E$ and $EM \sim E$ for any $n \times n$ matrix of polynomials E; for the same reason, $NE \sim E$ and $EN \sim E$. Thus $xI_n - A = xI_n - P^{-1}BP = P^{-1}(xI_n - B)P = ND^{-1}M(xI_n - B)M^{-1}DN^{-1} \sim D^{-1}M(xI_n - B)M^{-1}DN^{-1}$ (because N is unimodular) $\sim D^{-1}M(xI_n - B)M^{-1}D$ (because N^{-1} is unimodular) $\sim M(xI_n - B)M^{-1}$ (by what was just proved) $\sim xI_n - B$ (because M and M^{-1} are unimodular), as was to be shown.

Corollary. *If $A \simeq B$, $xI_n - A$ and $xI_n - B$ have the same determinant.*

For an $n \times n$ matrix A of rational numbers, the determinant of $xI_n - A$ is called the **characteristic polynomial** of A. Thus, the corollary says that similar matrices have the same characteristic polynomial.

The characteristic polynomial of an $n \times n$ matrix A is always monic of degree n, because the expansion of the determinant of $xI_n - A$ as a sum of $n!$ terms, each of which is plus or minus a product of n entries of $xI_n - A$, contains no terms of degree greater than n in x and only one term $(x - a_{11})(x - a_{22}) \cdots (x - a_{nn})$ of degree n; since this term is x^n plus terms of lower degree in x, the determinant of $xI_n - A$ is x^n plus terms of lower degree in x.

Similar matrices have the same characteristic polynomial but matrices with the same characteristic polynomial need not be similar. Consider the matrices

$$A = \begin{bmatrix} 0 & 0 & 0 \\ 0 & 0 & 0 \\ 0 & 0 & 0 \end{bmatrix} \quad \text{or} \quad \begin{bmatrix} 0 & 0 & 0 \\ 1 & 0 & 0 \\ 0 & 1 & 0 \end{bmatrix} \quad \text{or} \quad \begin{bmatrix} 0 & 0 & 0 \\ 0 & 0 & 0 \\ 0 & 1 & 0 \end{bmatrix}.$$

The strongly diagonal matrices equivalent to $xI_3 - A$ in these three cases are

$$\begin{bmatrix} x & 0 & 0 \\ 0 & x & 0 \\ 0 & 0 & x \end{bmatrix} \quad \text{or} \quad \begin{bmatrix} 1 & 0 & 0 \\ 0 & 1 & 0 \\ 0 & 0 & x^3 \end{bmatrix} \quad \text{or} \quad \begin{bmatrix} 1 & 0 & 0 \\ 0 & x & 0 \\ 0 & 0 & x^2 \end{bmatrix}$$

respectively (see the Examples section of Chapter 8). The characteristic polynomial is x^3 in all three cases, but, since these strongly diagonal matrices are not equal, they are not equivalent, and no two of the three matrices A are similar.

3 The Necessary Condition Is Sufficient

Theorem. *If A and B are $n \times n$ matrices of rational numbers, and if $xI - A \sim xI - B$, then $A \simeq B$.*

Proof. Since $xI_n - A$ and $xI_n - B$ are equivalent to each other, there exist unimodular matrices of polynomials M and N such that $xI - B = M(xI - A)N$. Let this equation be written in the form $(xI - B)N^{-1} = M(xI - A)$. There is a matrix of polynomials Q_1 and a matrix of rational numbers R_1 such that $M = (xI - B)Q_1 + R_1$. In fact, if $M = M_0x^k + M_1x^{k-1} + \cdots + M_k$, where the M_i are matrices of rational numbers, one can set $R_1 = B^kM_0 + B^{k-1}M_1 + \cdots + M_k$; then $M - R_1 = (x^kI - B^k)M_0 + (x^{k-1}I - B^{k-1})M_1 + \cdots + (xI - B)M_{k-1}$ is $xI - B$ times a matrix of polynomials because each term $(x^iI - B^i)M_{k-i}$ can be written $(xI - B)(x^{i-1}I + x^{i-2}B + \cdots + xB^{i-2} + B^{i-1})M_{k-i}$. Similarly, there is a matrix of polynomials Q_2 and a matrix of rational numbers R_2 such that $N^{-1} = Q_2(xI - A) + R_2$. (In fact, if $N^{-1} = L_0x^k + L_1x^{k-1} + \cdots$, then $R_2 = L_0A^k + L_1A^{k-1} + \cdots$.) Therefore, $(xI - B)(Q_2(xI - A) + R_2) = ((xI - B)Q_1 + R_1)(xI - A)$, which can be written in the form $(xI - B)(Q_2 - Q_1)(xI - A) = R_1(xI - A) - (xI - B)R_2$. The right side of this equation contains no entries of degree greater than 1, but if $Q_2 - Q_1$ were not zero, then the left side would contain entries of degree greater than 1 coming from the term of $xI(Q_2 - Q_1)xI = x^2(Q_2 - Q_1)$ of highest degree. Therefore, Q_2 and Q_1 must be equal and $R_1(xI - A)$ must equal $(xI - B)R_2$, that is, $xR_1 - R_1A = xR_2 - BR_2$. Therefore, $R_1 = R_2$ (equate coefficients of x) and $R_1A = BR_2 = BR_1$.

Let the above method be used to write M^{-1} in the form $(xI - A)Q_3 + R_3$, where R_3 is a matrix of numbers. Then $I = MM^{-1} = (xI - B)Q_1M^{-1} + R_1M^{-1} = (xI - B)Q_1M^{-1} + R_1(xI - A)Q_3 + R_1R_3 = (xI - B)Q_1M^{-1} + (xI - B)R_1Q_3 + R_1R_3 = (xI - B)(Q_1M^{-1} + R_1Q_3) + R_1R_3$. If $Q_1M^{-1} + R_1Q_3$ were not zero, the right side would contain terms in x, which it cannot because the left side is I. Therefore $I = R_1R_3$, so R_1 is invertible, and $A = R_1^{-1}R_1A = R_1^{-1}BR_1$. Therefore $A \simeq B$, as was to be shown.

This theorem makes it possible to determine whether two given matrices A and B are similar. Find strongly diagonal matrices equivalent to each of $xI - A$ and $xI - B$; then $A \simeq B$ if and only if these strongly diagonal matrices are the same.

4 Matrices Similar to Given Matrices

Theorem. *Given an $n \times n$ matrix of polynomials C whose determinant is monic of degree n, there is an $n \times n$ matrix of rational numbers A such that $xI - A \sim C$.*

Proof. Consider first the case in which C is a strongly diagonal matrix whose first $n-1$ diagonal entries are 1. In this case the last diagonal entry of C is monic of degree n, say it is $f(x) = x^n + a_1x^{n-1} + a_2x^{n-2} + \cdots + a_n$ where the a_i are rational numbers. An $n \times n$ matrix A satisfying $xI - A \sim C$ is found by taking the first $n-1$ columns of A to be the last $n-1$ columns of I_n and its last column to be the negatives of the coefficients of $f(x)$ in reverse order, that is, taking the entries of the last column of A, reading from top to bottom, to be $-a_n$, $-a_{n-1}$, $\ldots$, $-a_1$. (Note that the case $f(x) = x^3$ was treated above.) With this choice of A,

$$xI - A = \begin{bmatrix} x & 0 & 0 & \ldots & 0 & a_n \\ -1 & x & 0 & \ldots & 0 & a_{n-1} \\ 0 & -1 & x & \ldots & 0 & a_{n-2} \\ \vdots & \vdots & \vdots & \ddots & \vdots & \vdots \\ 0 & 0 & 0 & \ldots & x & a_2 \\ 0 & 0 & 0 & \ldots & -1 & x + a_1 \end{bmatrix}.$$

Let x times row i be added to row $i-1$ for $i = n, n-1, \ldots, 2$, in that order. These operations eliminate the x's in the first $n-1$ columns but make no other changes in these columns; in the last column, the last entry is unchanged, the next-to-last entry becomes $x^2+a_1x+a_2$, the entry above it becomes $x^3+a_1x^2+a_2x+a_3$, and so forth, and the top entry becomes $x^n + a_1x^{n-1} + a_2x^{n-2} + \cdots + a_{n-1}x + a_n = f(x)$. Let $g_i(x)$ denote the ith entry in the last column of the resulting matrix. Then

$$xI - A \sim \begin{bmatrix} 0 & 0 & 0 & \ldots & 0 & f(x) \\ -1 & 0 & 0 & \ldots & 0 & g_2(x) \\ 0 & -1 & 0 & \ldots & 0 & g_3(x) \\ \vdots & \vdots & \vdots & \ddots & \vdots & \vdots \\ 0 & 0 & 0 & \ldots & 0 & g_{n-1}(x) \\ 0 & 0 & 0 & \ldots & -1 & g_n(x) \end{bmatrix}.$$

Let D be the strongly diagonal matrix equivalent to $xI - A$. Since the matrix just given has an $(n-1) \times (n-1)$ minor equal to ± 1 (namely, the minor in the lower left corner) the gcd of its $(n-1) \times (n-1)$ minors must be 1, so the first $n-1$ diagonal entries of D must all be 1. On the other hand, the determinant of the matrix just given is, because $f(x)$ is the only nonzero entry in its row, $(-1)^{n+1} f(x)$ times the determinant of $-I_{n-1}$. Therefore, the determinant of D is $f(x)$, which implies that the last diagonal entry of D is $f(x)$, that is, $D = C$, as was to be shown.

Next let C be a diagonal matrix of polynomials whose diagonal entries are all monic. Let $f_1(x), f_2(x), \ldots, f_n(x)$ be the diagonal entries of C, and let d_i be the degree of $f_i(x)$. Since the determinant of C has degree n, $d_1+d_2+\cdots+d_n = n$. An $n \times n$ matrix A with $xI - A \sim C$ can be constructed as follows. For each i for which $d_i > 0$, let A_i be the $d_i \times d_i$ matrix constructed above with the property that $xI_{d_i} - A_i$ is equivalent to the diagonal matrix whose diagonal entries are $d_i - 1$ ones followed by $f_i(x)$, and let A be the $n \times n$ matrix that is zero in all locations except that it contains A_1 in the first d_1 rows of its first d_1 columns when $d_1 > 0$, A_2 in the next d_2 rows of its next d_2 columns when $d_2 > 0$, and so forth, ending with A_n in the last d_n rows of its last d_n columns when $d_n > 0$. (Note that the case in which the fs are x, x, x and the case in which they are 1, x, x^2 were treated above. Note also that the case in which all fs but the last are 1 is the case previously treated.) This matrix A can be seen to have the required property $xI - A \sim C$ as follows.

Let i be the smallest integer for which $d_i > 0$, and let row and column operations that transform $xI_{d_i} - A_i$ into the diagonal matrix with diagonal entries $1, 1, \ldots, 1, f_i(x)$ be applied to the first d_i rows of the first d_i columns of $xI_n - A$. They transform the upper left $d_i \times d_i$ corner of this matrix into the diagonal matrix with entries $1, 1, \ldots, 1, f_i(x)$ without changing any other entries of $xI_n - A$, because the affected rows are all zero in columns after column d_i and the affected columns are zero in rows after row d_i. In the same way, if d_j is the next one of the ds that is positive, the matrix in the next d_j rows of the next d_j columns can be transformed into the diagonal matrix with diagonal entries $1, 1, \ldots, 1, f_j(x)$ without changing any other entries. Continuation of this process transforms $xI_n - A$ into an equivalent diagonal matrix whose diagonal entries are the polynomials $f_i(x)$ of positive degree and ones. Thus, the diagonal entries of the diagonal matrix that is obtained are the same as the diagonal entries of C, but they are likely to be in a different order. Because

$$\begin{bmatrix} a & 0 \\ 0 & b \end{bmatrix} \sim \begin{bmatrix} 0 & -b \\ a & 0 \end{bmatrix} \sim \begin{bmatrix} b & 0 \\ 0 & a \end{bmatrix}$$

(multiply first on the left by the unimodular matrix $\begin{bmatrix} 0 & -1 \\ 1 & 0 \end{bmatrix}$ and then on the right by its inverse $\begin{bmatrix} 0 & 1 \\ -1 & 0 \end{bmatrix}$), a diagonal matrix is equivalent to any matrix obtained by rearranging its diagonal entries. Therefore, $xI_n - A \sim C$.

Finally, if C is any $n \times n$ matrix of polynomials whose determinant is monic of degree n, then C is equivalent to a strongly diagonal matrix of polynomials D. Since C and D have the same determinant, since $|C|$ is monic of degree n, and since all diagonal entries of D other than the last are monic, all diagonal entries of D are monic and their product has degree n. Therefore, by what was just shown, there is an $n \times n$ matrix of rational numbers A with $xI - A \sim D$. Then $xI - A \sim C$ and the proof is complete.

5 Rational Canonical Form

The method of the preceding section gives a solution to the problem of finding a simple matrix similar to a given one. Let B be a given $n \times n$ matrix of rational numbers. First, find a diagonal matrix of polynomials C, with monic diagonal entries, such that $C \sim xI - B$. Then, for each diagonal entry $f_i(x)$ of C of degree greater than 0, find a matrix A_i such that $xI - A_i$ is equivalent to the diagonal matrix whose last diagonal entry is $f_i(x)$ and whose remaining diagonal entries are all 1. Finally, set A equal to the matrix that contains $A_1, A_2, \ldots$ as "blocks" down the diagonal and is zero elsewhere. (Specifically, let A_i be $n_i \times n_i$. Then the sum of the n_i is n and the entry a_{rs} of A in row r and column s is zero unless, for some i, both r and s are greater than $S_i = n_1 + n_2 + \cdots + n_{i-1}$ and less than or equal to $S_i + n_i$; when there is such an i, a_{rs} is the entry in row $r - S_i$ and column $s - S_i$ of A_i.) The resulting matrix A satisfies $xI - A \sim C$, as was seen in the preceding section, so $A \simeq B$ by the theorem of Section 3. A is simple in the sense that it normally has many zero entries.

The choices in this construction—the choice of a diagonal C and the choice of an A_i for each diagonal entry $f_i(x) \neq 1$ of the chosen C—are always possible. For example, C can be taken to be the unique strongly diagonal matrix equivalent to $xI - B$, and A_i can be taken to be the matrix of this type described in the last section. Different choices lead to a different As. A standard way to make the choices, which leads to what is called the **rational canonical form** of B, is to base them on two propositions:

Proposition 1. *For polynomials $f(x)$ and $g(x)$, the matrices*

$$\begin{bmatrix} f(x) & 0 \\ 0 & g(x) \end{bmatrix} \quad \textit{and} \quad \begin{bmatrix} 1 & 0 \\ 0 & f(x)g(x) \end{bmatrix}$$

are equivalent if and only if $f(x)$ and $g(x)$ are relatively prime, that is, if and only if the greatest common divisor of $f(x)$ and $g(x)$ is 1.

Proof. Let M_1 be the first of these two matrices, and let M_2 be the second. Let D be the strongly diagonal matrix equivalent to M_1. Since M_2 is strongly diagonal, $M_1 \sim M_2$ if and only if $M_2 = D$. Let $d(x)$ be the first diagonal entry of D.

Then $d(x)$ is monic, $d(x)$ divides both $f(x)$ and $g(x)$ (because it divides all entries of a matrix equivalent to M_1), and $d(x)$ is divisible by any polynomial that divides both $f(x)$ and $g(x)$ (because such a polynomial divides all entries of M_1 and therefore of D). In short, $d(x)$ is the greatest common divisor of $f(x)$ and $g(x)$. Since the determinant of D is $f(x)g(x)$, the other diagonal entry of D is $f(x)g(x)/d(x)$. Therefore, $D = M_2$ if and only if $d(x) = 1$, as was to be shown.

The diagonal entries of a diagonal matrix C equivalent to $xI - B$ give a factorization of the characteristic polynomial of B. The proposition shows that if one of the diagonal entries of C can be written as a product of two relatively prime factors, then there is another diagonal matrix equivalent to $xI - B$ whose diagonal entries give a new factorization of the characteristic polynomial of B into more factors. Starting with the *strongly* diagonal matrix C equivalent to $xI - B$ and repeatedly applying this reduction* step—always taking *monic* factors—one arrives eventually at a diagonal matrix C equivalent to $xI - B$ in which the diagonal entries are monic polynomials, none of which can be decomposed as products of two relatively prime factors. This matrix is determined, except for the order of its diagonal entries, by B (see Exercise 13). Its diagonal entries of positive degree are called the **elementary divisors** of B. (Note that the elementary divisors of B are not divisors of B at all; they are divisors of the characteristic polynomial of B.)

The rational canonical form of B is the matrix A similar to B given by the construction described above when, for each elementary divisor of B, the $n_i \times n_i$ block described by Proposition 2 is used. Proposition 2 applies to all elementary divisors, because a monic polynomial which cannot be written as a product of two relatively prime factors of lower degree must be a power of an irreducible monic polynomial (see Exercise 12).

Proposition 2. *Let $f(x) = x^n + a_1x^{n-1} + \cdots + a_n$ be an irreducible polynomial, and let A be the $n \times n$ matrix whose first $n - 1$ columns are the last $n - 1$ columns of I_n and whose last column contains $-a_n, -a_{n-1}, \ldots, -a_1$, in that order. The strongly diagonal matrix equivalent to $xI - A$ has diagonal entries* 1, 1, ..., 1, *$f(x)$, as was shown in the last section. For $j > 1$, let $A^{(j)}$ be the $nj \times nj$ matrix*

*The reduction requires that polynomials be factored. The factorization of polynomials is very like the factorization of integers—fairly easy for small cases, but of rapidly increasing difficulty as the size of the object to be factored increases. We will ignore the technical problem of factoring polynomials and will take for granted that every monic polynomial can be decomposed into monic *irreduible* factors, that is, factors which can not be written as products of polynomials of lower degree.

of rational numbers given schematically by the formula

$$A^{(j)} = \begin{bmatrix} A & 0 & 0 & \cdots & 0 & 0 \\ I & A & 0 & \cdots & 0 & 0 \\ 0 & I & A & \cdots & 0 & 0 \\ \vdots & \vdots & \vdots & \ddots & \vdots & \vdots \\ 0 & 0 & 0 & \cdots & A & 0 \\ 0 & 0 & 0 & \cdots & I & A \end{bmatrix}.$$

(Thus, rows $ni + 1$ through $ni + n$ of columns $ni + 1$ through $ni + n$ contain A for $i = 0, 1, \ldots, j - 1$; rows $ni + 1$ through $ni + n$ of columns $n(i - 1) + 1$ through ni contain I_n for $i = 1, 2, \ldots, j - 1$; and all other entries are zero.) Then $xI - A^{(j)}$ is equivalent to the diagonal matrix whose diagonal entries are $1, 1, \ldots, 1, f(x)^j$.

For the proof, see Exercise 18.

In summary, the construction of the rational canonical form of B is as follows. Find the elementary divisors of B and arrange them in some order, say they are $e_1(x), e_2(x), \ldots, e_k(x)$. (To be precise, the rational canonical form is not truly canonical until a rule is established for ordering the elementary divisors. This is a matter of no importance.) For each $e_i(x)$, let A_i be the matrix with $xI - A_i$ equivalent to $1, 1, \ldots, 1, e_i(x)$ that is given by Proposition 2. The rational canonical form A of B is the matrix which has the matrices $A_1, A_2, \ldots, A_k$ in blocks down the diagonal in the manner described in the first paragraph of this section.

6 The Minimum Polynomial of a Matrix

A square matrix A can be multiplied by itself any number of times, which is to say that it can be raised to any power. More generally, since an $n \times n$ matrix can also be multiplied by a rational number—meaning, as always, that every entry is multiplied by that rational number—there is a natural meaning of $f(A)$ for any polynomial $f(x)$ with rational coefficients, namely, for $f(x) = a_0x^n + a_1x^{n-1} + \cdots + a_n$, $f(A)$ means the $n \times n$ matrix $a_nA^n + a_{n-1}A^{n-1} + \cdots + a_nI_n$. Because the same rules of arithmetic apply in the two cases—namely, the associative laws of addition and multiplication, the distributive law, the commutative law of addition, and the laws $x^mx^n = x^{m+n}$ and $A^mA^n = A^{m+n}$—if $f(x)$, $g(x)$, $h(x)$ are polynomials such that $f(x) = g(x)h(x)$, and if A is a square matrix, then $f(A) = g(A)h(A)$.

If A is a square matrix of rational numbers and if $f(x)$ is a polynomial with rational coefficients, we will say that A is a **root** of $f(x)$ if $f(A) = 0$, that is, if the matrix $f(A)$ has all entries equal to zero. Given any A, not only are there polynomials of which A is a root, there is a unique monic polynomial with the property that the polynomials of which A is a root are the multiples of this one. This polynomial, which is called the **minimum polynomial** of A, is given specifically by:

Theorem. *Given a square matrix of rational numbers A, let $f(x)$ be the last diagonal entry of the strongly diagonal matrix equivalent to $xI - A$. Then A is a root of a polynomial $g(x)$ with rational coefficients if and only if $g(x)$ is a multiple of $f(x)$.*

(If A is a 1×1 matrix, the theorem is simply the familar statement that a rational number a is a root of a polynomial $g(x)$ if and only if $g(x)$ is a multiple of $x - a$.)

Proof. Consider first the case in which A is an $n \times n$ matrix whose first $n - 1$ columns are the last $n - 1$ columns of I_n. Another way to state this condition is to say that $Ae_i = e_{i+1}$ for $i = 1, 2, \ldots, n - 1$, where e_i denotes the ith column of I_n. It was shown in Section 4 that the strongly diagonal matrix equivalent to $xI - A$ in this case has as its diagonal entries $1, 1, \ldots, 1, f(x)$, where $f(x)$ is monic of degree n. Moreover, the coefficients of $f(x) = x^n + a_1x^{n-1} + \cdots + a_n$ are the negatives of the entries in the last column of A in reverse order; that is, a_1 is the negative of the entry in the lower right corner of A, a_2 is the negative of the entry above that one, and so forth. Now $A^n e_1 = A^{n-1}e_2 = A^{n-2}e_3 = \cdots = Ae_n$ is the last column of A, which is $-a_ne_1 - a_{n-1}e_2 - \cdots - a_1e_n = -a_nIe_1 - a_{n-1}Ae_1 - \cdots - a_1A^{n-1}e_1$. Thus, $f(A)e_1 = 0$. Then for any column e_i of I_n, the formula $f(A)e_i = f(A)A^{i-1}e_1 = A^{i-1}f(A)e_1 = A^{i-1}0 = 0$ holds. Therefore, $f(A) = f(A)I = 0$, so $g(A) = 0$ whenever $g(x)$ is a multiple of $f(x)$. Now let $g(x)$ be any polynomial of which A is a root, and let $d(x)$ be the greatest common divisor of $f(x)$ and $g(x)$. Then $d(x)$ is monic, $d(x) = r(x)f(x) + s(x)g(x)$ for some polynomials $r(x)$ and $s(x)$, and $f(x)$ and $g(x)$ are both multiples of $d(x)$. Since $d(A) = r(A)f(A) + s(A)g(A) = r(A) \cdot 0 + s(A) \cdot 0 = 0$, the degree of the monic polynomial $d(x)$ is at least 1. Since $f(x)$ is a multiple of $d(x)$, the degree of $d(x)$ is at most n, say $d(x) = d_0x^n + d_1x^{n-1} + \cdots + d_n$. If d_0 were zero then $d(A) = 0$ would imply $(d_1A^{n-1} + d_2A^{n-2} + \cdots + d_nI)e_1 = 0$, that is, $d_1e_n + d_2e_{n-1} + \cdots + d_ne_1 = 0$, which would imply that all the ds were 0, contrary to the fact that $d(x)$ is monic. Therefore, $d(x)$ has degree n. Since $f(x)$ is a multiple of $d(x)$ and both are monic, it follows that $d(x) = f(x)$. Therefore,

$g(x)$ is a multiple of $f(x)$. In short, A is a root of $g(x)$ if and only if $g(x)$ is a multiple of $f(x)$, as was to be shown.

Next let A be a matrix obtained by applying the construction of Section 4 to a strongly diagonal $n \times n$ matrix of polynomials D whose determinant is monic of degree n. Specifically, let $f_1(x), f_2(x), \ldots, f_k(x)$ be the diagonal entries of D, in order, that have positive degree. Let n_i be the degree of $f_i(x)$, and let A_i be the matrix that is related to $f_i(x)$ in the way that A is related to $f(x)$ in the preceding paragraph. Then A is the $n \times n$ matrix of rational numbers that has $A_1, A_2, \ldots, A_k$ in blocks down the diagonal as in Sections 4 and 5. It follows directly from the definition of matrix multiplication that A^2 has $A_1^2, A_2^2, \ldots, A_k^2$ in blocks down the diagonal and that the analogous rule holds for any power of A. Therefore, for any polynomial $g(x)$, $g(A)$ has $g(A_1), g(A_2), \ldots, g(A_k)$ in blocks down the diagonal. Consequently, $g(A) = 0$ if and only if $g(A_i) = 0$ for each i. As was just shown, this is true if and only if $g(x)$ is a multiple of $f_i(x)$ for each i. But D is strongly diagonal, so $g(x)$ is a multiple of $f_i(x)$ for each i if and only if it is a multiple of $f_k(x)$. In short, $g(A) = 0$ if and only if $g(x)$ is a multiple of $f_k(x)$, as was to be shown.

Finally, let A be an arbitrary $n \times n$ matrix of rational numbers. Let D be the strongly diagonal matrix equivalent to $xI - A$, let $f_1(x), f_2(x), \ldots, f_k(x)$ be its entries of positive degree, and let B be the matrix obtained when the construction of Section 4 is applied to D. As has just been shown, $g(B) = 0$ if and only if $g(x)$ is a multiple of $f_k(x)$. By the theorem of Section 3, $A \simeq B$, that is, there is an invertible matrix P such that $A = P^{-1}BP$. For any polynomial $g(x)$ with rational coefficients, say $g(x) = b_0x^m + b_1x^{m-1} + \cdots + b_m$, $g(A) = b_0A^m + b_1A^{m-1} + \cdots + b_mI = b_0(P^{-1}BP)^m + b_1(P^{-1}BP)^{m-1} + \cdots + b_mI = b_0P^{-1}B^mP + b_1P^{-1}B^{m-1}P + \cdots + b_mP^{-1}IP = P^{-1}(b_0B^m + b_1B^{m-1} + \cdots + b_mI)P = P^{-1}g(B)P$. Therefore, $g(A) = 0$ if and only if $g(B) = 0$, which is true if and only if $g(x)$ is a multiple of $f_k(x)$, as was to be shown.

Note that similar matrices have the same minimum polynomial.

Corollary (Cayley-Hamilton Theorem). *A square matrix of rational numbers is a root of its characteristic polynomial.*

Proof. Because $F(x)$ is the determinant of the strongly diagonal matrix D equivalent to $xI - A$, and because the minimum polynomial of A is a diagonal entry of D, $F(x)$ is a multiple of the minimum polynomial of A. Therefore, $F(A) = 0$.

7 Diagonalizable Matrices

An $n \times n$ matrix of rational numbers A is **diagonalizable** if it is similar to a diagonal matrix.

Proposition. *A matrix is diagonalizable if and only if its minimum polynomial is of the form* $(x - a_1)(x - a_2) \cdots (x - a_l)$, *where* $a_1, a_2, \ldots, a_l$ *are distinct rational numbers.*

Proof. Let $f(x)$ be the minimum polynomial of a square matrix A.

First suppose A is diagonalizable, say it is similar to the diagonal matrix B. Let $b_1, b_2, \ldots, b_l$ be a list of the *distinct* diagonal entries of B, and let $h(x) = (x - b_1)(x - b_2) \cdots (x - b_l)$. Since $h(B)$ is a diagonal matrix in which the diagonal entries all have the form $h(b_i)$, $h(B) = 0$. Therefore, $h(x)$ is a multiple of $f(x)$ ($f(x)$ is the minimum polynomial of B because similar matrices have the same minimum polynomial), say $h(x) = f(x)q(x)$. On the other hand, since the diagonal entries of $f(B)$ are all zero, $f(x)$ has l distinct roots $b_1, b_2, \ldots, b_l$, so the degree of $f(x)$ is at least $l = \deg h(x)$ (see Exercise 11). Since $f(x)$ and $h(x)$ are both monic, it follows that $q(x) = 1$ and that $f(x) = h(x)$. Therefore, $f(x)$ has the required form.

Conversely, suppose $f(x) = (x - a_1)(x - a_2) \cdots (x - a_l)$, where $a_1, a_2, \ldots, a_l$ are distinct rational numbers. An elementary divisor $e(x)$ of A is by definition a factor of one of the diagonal entries of the strongly diagonal matrix equivalent to $xI - A$. Therefore, since $f(x)$ is a multiple of each of these diagonal entries, $f(x)$ is a multiple of $e(x)$, say $f(x) = e(x)q(x)$. Each a_i is a root of $f(x)$ and is therefore a root either of $e(x)$ or of $q(x)$. At most one a_i can be a root of $e(x)$, because the elementary divisors of A are powers of irreducible polynomials (Exercise12). Therefore, $q(x)$ has at least $l - 1$ distinct roots, which implies $\deg q(x) \geq l - 1$ and $\deg e(x) = \deg f(x) - \deg q(x) \leq 1$. Because an elementary divisor is monic and has positive degree, it follows that $e(x) = x - a_i$ for some i. Since this is true for all elementary divisors of A, the rational canonical form of A is a diagonal matrix. Therefore, A is diagonalizable, as was to be shown.

8 Other Coefficient Fields

The definitions and theorems of this chapter apply equally well to matrices of other sorts of numbers, such as real numbers or complex numbers. In fact, the proofs that have been given apply in these other cases without change.

If numbers other than rational numbers are allowed, the meaning of "irreducible" changes, which implies that the elementary divisors of a matrix may change and consequently that the rational canonical form of a matrix may change. For example, when complex numbers are allowed, the rational canonical form becomes what is called the **Jordan canonical form**. Because *the only monic irreducible polynomials with complex coefficients are the polynomials* $x - a$ *where* a *is a complex number,* the elementary divisors of a square matrix over the complex numbers are all of the form $(x - a)^j$ and the blocks in the rational (Jordan) canonical form of a matrix are either 1×1 or of the form

$$\begin{bmatrix} a & 0 & 0 & \cdots & 0 & 0 \\ 1 & a & 0 & \cdots & 0 & 0 \\ 0 & 1 & a & \cdots & 0 & 0 \\ \vdots & \vdots & \vdots & \ddots & \vdots & \vdots \\ 0 & 0 & 0 & \cdots & a & 0 \\ 0 & 0 & 0 & \cdots & 1 & a \end{bmatrix}$$

for some complex number a.

Examples

Let $A = \begin{bmatrix} 1 & 0 \\ 0 & 2 \end{bmatrix}$. A matrix similar to A can be constructed by choosing an invertible matrix P and computing PAP^{-1}. For example, with $P = \begin{bmatrix} 3 & 1 \\ 5 & 3 \end{bmatrix}$ (invertible because its determinant is nonzero) one finds

$$B = PAP^{-1} = \begin{bmatrix} 3 & 1 \\ 5 & 3 \end{bmatrix}\begin{bmatrix} 1 & 0 \\ 0 & 2 \end{bmatrix}\begin{bmatrix} \frac{3}{4} & -\frac{1}{4} \\ -\frac{5}{4} & \frac{3}{4} \end{bmatrix} = \begin{bmatrix} -\frac{1}{4} & \frac{3}{4} \\ -\frac{15}{4} & \frac{13}{4} \end{bmatrix}.$$

The theorem of Section 2 says that $xI - A$ and $xI - B$ must be equivalent as matrices of polynomials. Indeed,

$$\begin{aligned} xI - A &= \begin{bmatrix} x-1 & 0 \\ 0 & x-2 \end{bmatrix} \sim \begin{bmatrix} x-1 & x-2 \\ 0 & x-2 \end{bmatrix} \sim \begin{bmatrix} x-1 & -1 \\ 0 & x-2 \end{bmatrix} \sim \\ &\sim \begin{bmatrix} -1 & -1 \\ x^2-2x & x-2 \end{bmatrix} \sim \begin{bmatrix} 1 & -1 \\ x^2-4x+4 & x-2 \end{bmatrix} \sim \\ &\sim \begin{bmatrix} 1 & 0 \\ x^2-4x+4 & x^2-3x+2 \end{bmatrix} \sim \begin{bmatrix} 1 & 0 \\ 0 & x^2-3x+2 \end{bmatrix} \end{aligned}$$

and

$$\begin{aligned} xI - B &= \begin{bmatrix} x+\frac{1}{4} & -\frac{3}{4} \\ \frac{15}{4} & x-\frac{13}{4} \end{bmatrix} \sim \begin{bmatrix} \frac{1}{4} & -\frac{3}{4} \\ \frac{4}{3}x^2 - \frac{13}{3}x + \frac{15}{4} & x-\frac{13}{4} \end{bmatrix} \sim \\ &\sim \begin{bmatrix} 1 & -\frac{3}{4} \\ \frac{4}{3}x^2 - \frac{16}{3}x + 7 & x-\frac{13}{4} \end{bmatrix} \sim \begin{bmatrix} 1 & 0 \\ \frac{4}{3}x^2 - \frac{16}{3}x + 7 & x^2-3x+2 \end{bmatrix} \\ &\sim \begin{bmatrix} 1 & 0 \\ 0 & x^2-3x+2 \end{bmatrix}, \end{aligned}$$

so $xI - A \sim xI - B$. Since A is diagonal, $g(A) = \begin{bmatrix} g(1) & 0 \\ 0 & g(2) \end{bmatrix}$ for any polynomial $g(x)$. Therefore, the minimum polynomial of A is the polynomial $(x-1)(x-2) = x^2 - 3x + 2$ with roots 1 and 2. That $x^2 - 3x + 2$ is also the minimum polynomial of B (which it must be because $B \simeq A$) is less obvious. However,

$$B^2 - 3B + 2I = \begin{bmatrix} -\frac{1}{4} & \frac{3}{4} \\ -\frac{15}{4} & \frac{13}{4} \end{bmatrix}\begin{bmatrix} -\frac{1}{4} & \frac{3}{4} \\ -\frac{15}{4} & \frac{13}{4} \end{bmatrix} - 3\begin{bmatrix} -\frac{1}{4} & \frac{3}{4} \\ -\frac{15}{4} & \frac{13}{4} \end{bmatrix} + \begin{bmatrix} 2 & 0 \\ 0 & 2 \end{bmatrix}$$

is easily computed and found to be zero; since B obviously satisfies no equation of the form $B - cI = 0$, $x^2 - 3x + 2$ is the minimum polynomial of B. The rational canonical form of these matrices is found by first doing the above computation of $xI - A \sim \begin{bmatrix} 1 & 0 \\ 0 & x^2 - 3x + 2 \end{bmatrix}$ and factoring $x^2 - 3x + 2 = (x-1)(x-2)$ to find that the elementary divisors of A (or B) are $x - 1$ and $x - 2$. Therefore, A is itself in rational canonical form, and A is the rational canonical form of B.

The proof of Section 3 gives a different matrix P with $B = PAP^{-1}$ than the one used to define B above. The above computations show that $D = \begin{bmatrix} 1 & 0 \\ 0 & x^2 - 3x + 2 \end{bmatrix}$ can be written as

$$\begin{bmatrix} 1 & 0 \\ -x^2 + 4x - 4 & 1 \end{bmatrix}\begin{bmatrix} 1 & 1 \\ 0 & 1 \end{bmatrix}(xI - A)\begin{bmatrix} 1 & -1 \\ 0 & 1 \end{bmatrix}\begin{bmatrix} 1 & 0 \\ x & 1 \end{bmatrix}\begin{bmatrix} 1 & 0 \\ -2 & 1 \end{bmatrix}\begin{bmatrix} 1 & 1 \\ 0 & 1 \end{bmatrix}$$

and as

$$\begin{bmatrix} 1 & 0 \\ -\frac{4}{3}x^2 + \frac{16}{3}x - 7 & 1 \end{bmatrix}(xI - B)\begin{bmatrix} 1 & 0 \\ \frac{4}{3}x & 1 \end{bmatrix}\begin{bmatrix} 1 & 0 \\ -1 & 1 \end{bmatrix}\begin{bmatrix} 1 & \frac{3}{4} \\ 0 & 1 \end{bmatrix}.$$

Therefore, $xI - B = M(xI - A)N$ where

$$\begin{aligned} M &= \begin{bmatrix} 1 & 0 \\ \frac{4}{3}x^2 - \frac{16}{3}x + 7 & 1 \end{bmatrix}\begin{bmatrix} 1 & 0 \\ -x^2 + 4x - 4 & 1 \end{bmatrix}\begin{bmatrix} 1 & 1 \\ 0 & 1 \end{bmatrix} \\ &= \begin{bmatrix} 1 & 1 \\ \frac{1}{3}x^2 - \frac{4}{3}x + 3 & \frac{1}{3}x^2 - \frac{4}{3}x + 4 \end{bmatrix}. \end{aligned}$$

(One can also find

$$\begin{aligned} N &= \begin{bmatrix} 1 & -1 \\ 0 & 1 \end{bmatrix}\begin{bmatrix} 1 & 0 \\ x & 1 \end{bmatrix}\begin{bmatrix} 1 & 0 \\ -2 & 1 \end{bmatrix}\begin{bmatrix} 1 & 1 \\ 0 & 1 \end{bmatrix}\begin{bmatrix} 1 & -\frac{3}{4} \\ 0 & 1 \end{bmatrix}\begin{bmatrix} 1 & 0 \\ 1 & 1 \end{bmatrix}\begin{bmatrix} 1 & 0 \\ -\frac{4}{3}x & 1 \end{bmatrix} \\ &= \begin{bmatrix} \frac{1}{3}x^2 - \frac{11}{12}x + \frac{11}{4} & -\frac{1}{4}x - \frac{1}{4} \\ -\frac{1}{3}x^2 + \frac{7}{12}x - \frac{3}{2} & \frac{1}{4}x + \frac{1}{2} \end{bmatrix}, \end{aligned}$$

but N will not be needed.) In the notation of Section 3, $A = R_1^{-1}BR_1$ where

$$\begin{aligned} R_1 &= B^2M_0 + BM_1 + M_2 \\ &= \begin{bmatrix} -\frac{11}{4} & \frac{9}{4} \\ -\frac{45}{4} & \frac{31}{4} \end{bmatrix}\begin{bmatrix} 0 & 0 \\ \frac{1}{3} & \frac{1}{3} \end{bmatrix} + \begin{bmatrix} -\frac{1}{4} & \frac{3}{4} \\ -\frac{15}{4} & \frac{13}{4} \end{bmatrix}\begin{bmatrix} 0 & 0 \\ -\frac{4}{3} & -\frac{4}{3} \end{bmatrix} + \begin{bmatrix} 1 & 1 \\ 3 & 4 \end{bmatrix} \\ &= \begin{bmatrix} \frac{3}{4} & \frac{3}{4} \\ \frac{5}{4} & \frac{9}{4} \end{bmatrix}. \end{aligned}$$

It is easily checked that $B = R_1AR_1^{-1}$.

Consider next the matrix $A = \begin{bmatrix} -1 & 1 & -1 \\ -9 & 2 & -3 \\ 0 & -3 & 2 \end{bmatrix}$. Its minimum polynomial can be found by reducing $xI - A$ to strongly diagonal form.

$$xI - A = \begin{bmatrix} x+1 & -1 & 1 \\ 9 & x-2 & 3 \\ 0 & 3 & x-2 \end{bmatrix} \sim \begin{bmatrix} x+1 & 1 & 1 \\ 9 & x+4 & 3 \\ 0 & 2x-1 & x-2 \end{bmatrix} \sim$$

$$\sim \begin{bmatrix} x+1 & 1 & 0 \\ 9 & x+4 & -x-1 \\ 0 & 2x-1 & -x-1 \end{bmatrix} \sim \begin{bmatrix} 1 & 1 & 0 \\ -x^2-4x+9 & x+4 & -x-1 \\ -2x^2+x & 2x-1 & -x-1 \end{bmatrix} \sim$$

$$\sim \begin{bmatrix} 1 & 0 & 0 \\ -x^2-4x+9 & x^2+5x-5 & -x-1 \\ -2x^2+x & 2x^2+x-1 & -x-1 \end{bmatrix} \sim \begin{bmatrix} 1 & 0 & 0 \\ 0 & x^2+5x-5 & -x-1 \\ 0 & 2x^2+x-1 & -x-1 \end{bmatrix} \sim$$

$$\sim \begin{bmatrix} 1 & 0 & 0 \\ 0 & 4x-5 & -x-1 \\ 0 & x^2-1 & -x-1 \end{bmatrix} \sim \begin{bmatrix} 1 & 0 & 0 \\ 0 & x-8 & -x-1 \\ 0 & x^2-3x-4 & -x-1 \end{bmatrix} \sim$$

$$\sim \begin{bmatrix} 1 & 0 & 0 \\ 0 & x-8 & -9 \\ 0 & x^2-3x-4 & x^2-4x-5 \end{bmatrix} \sim$$

$$\sim \begin{bmatrix} 1 & 0 & 0 \\ 0 & -8 & -9 \\ 0 & \frac{1}{9}x^3+\frac{5}{9}x^2-\frac{32}{9}x-4 & x^2-4x-5 \end{bmatrix} \sim$$

$$\sim \begin{bmatrix} 1 & 0 & 0 \\ 0 & 1 & -9 \\ 0 & \frac{1}{9}x^3-\frac{4}{9}x^2+\frac{4}{9}x+1 & x^2-4x-5 \end{bmatrix} \sim$$

$$\sim \begin{bmatrix} 1 & 0 & 0 \\ 0 & 1 & 0 \\ 0 & \frac{1}{9}x^3-\frac{4}{9}x^2+\frac{4}{9}x+1 & x^3-3x^2+4 \end{bmatrix} \sim \begin{bmatrix} 1 & 0 & 0 \\ 0 & 1 & 0 \\ 0 & 0 & x^3-3x^2+4 \end{bmatrix}.$$

(Alternatively, the 2×2 minor in the lower left corner of $xI - Ax$ is 27, so the gcd of the 2×2 minors is 1; this shows that the first two diagonal entries of the equivalent strongly diagonal matrix are 1, after which the determination of the whole matrix is reduced to the computation of $|xI - A|$.) The minimum polynomial of A is therefore $x^3-3x^2+4 = (x+1)(x^2-4x+4) = (x+1)(x-2)^2$. That A is a "root" of this polynomial can be shown by computing $A + I = \begin{bmatrix} 0 & 1 & -1 \\ -9 & 3 & -3 \\ 0 & -3 & 3 \end{bmatrix}$, $A - 2I = \begin{bmatrix} -3 & 1 & -1 \\ -9 & 0 & -3 \\ 0 & -3 & 0 \end{bmatrix}$, $(A-2I)^2 = \begin{bmatrix} 0 & 0 & 0 \\ 27 & 0 & 9 \\ 27 & 0 & 9 \end{bmatrix}$, and $(A+I)(A-2I)^2 = 0$. Therefore, the minimum polynomial must divide $(x+1)(x-2)^2$, which means that it is this polynomial unless it is a proper factor of this polynomial; but $(A-2I)^2$ was just shown to be nonzero, and every other proper factor of $(x+1)(x-2)^2$ divides $(x+1)(x-2)$ and $(A+I)(A-2I) \neq 0$, so $(x+1)(x-2)^2$ is indeed the minimum polynomial of A.

Because the minimum polynomial of A is not a product of distinct linear factors, A is not diagonalizable. The elementary divisors of A are $x+1$ and $(x-2)^2$, and its rational canonical form is $\begin{bmatrix} -1 & 0 & 0 \\ 0 & 2 & 0 \\ 0 & 1 & 2 \end{bmatrix}$.

The characteristic polynomial of $A = \begin{bmatrix} 0 & 1 \\ -1 & 0 \end{bmatrix}$ is $\begin{vmatrix} x & -1 \\ 1 & x \end{vmatrix} = x^2+1$. Since the minimum polynomial of A is a factor of its characteristic polynomial, and since the characteristic polynomial is irreducible, x^2+1 must also be the minimum polynomial of A. (Alternatively, the minimum polynomial of A can be found by the computation $xI - A = \begin{bmatrix} x & -1 \\ 1 & x \end{bmatrix} \sim \begin{bmatrix} -1 & -1 \\ x^2+x+1 & x \end{bmatrix} \sim \begin{bmatrix} 1 & -1 \\ x^2-x+1 & x \end{bmatrix} \sim \begin{bmatrix} 1 & 0 \\ x^2-x+1 & x^2+1 \end{bmatrix} \sim \begin{bmatrix} 1 & 0 \\ 0 & x^2+1 \end{bmatrix}$.) The irreducible polynomial x^2+1 is the sole elementary divisor of A and the rational canonical form of A is $\begin{bmatrix} 0 & -1 \\ 1 & 0 \end{bmatrix}$, that is, the rational canonical form of A is $-A$. That A is similar to $-A$ can be proved directly by using the above reduction of $xI - A$ to strongly diagonal form and the analogous reduction of $xI + A$ to strongly diagonal form to find that $xI - A = M(xI + A)N$, where

$$M = \begin{bmatrix} 1 & 0 \\ x^2-x+1 & 1 \end{bmatrix}^{-1} \begin{bmatrix} 1 & 0 \\ -x^2+x-1 & 1 \end{bmatrix} = \begin{bmatrix} 1 & 0 \\ -2x^2+2x-2 & 1 \end{bmatrix}$$

and then applying the formulas of Section 3 to find

$$\begin{aligned} R_1 &= (-A)^2\begin{bmatrix} 0 & 0 \\ -2 & 0 \end{bmatrix} + (-A)\begin{bmatrix} 0 & 0 \\ 2 & 0 \end{bmatrix} + \begin{bmatrix} 1 & 0 \\ -2 & 1 \end{bmatrix} \\ &= \begin{bmatrix} 0 & 0 \\ 2 & 0 \end{bmatrix} + \begin{bmatrix} -2 & 0 \\ 0 & 0 \end{bmatrix} + \begin{bmatrix} 1 & 0 \\ -2 & 1 \end{bmatrix} \\ &= \begin{bmatrix} -1 & 0 \\ 0 & 1 \end{bmatrix} \end{aligned}$$

as a matrix for which $A = R_1^{-1}(-A)R_1$. (Given two similar matrices B and C, the problem of finding a matrix P that *demonstrates* that they are similar—that is, a matrix P that satisfies $B = P^{-1}CP$—is normally fairly difficult and cannot be solved by inspection. In the simple case of this example $B = \begin{bmatrix} 0 & 1 \\ -1 & 0 \end{bmatrix}$, $C = \begin{bmatrix} 0 & -1 \\ 1 & 0 \end{bmatrix}$, however, one can imagine guessing the solution $P = \begin{bmatrix} -1 & 0 \\ 0 & 1 \end{bmatrix}$ or the solution $P = \begin{bmatrix} 1 & 0 \\ 0 & -1 \end{bmatrix}$, at least with the benefit of hindsight.) If complex numbers are allowed, the minimum polynomial x^2+1 has the factorization $(x-i)(x+i)$, where $i = \sqrt{-1}$, and the Jordan canonical form of A is the diagonal matrix $\begin{bmatrix} i & 0 \\ 0 & -i \end{bmatrix}$. For a matrix P with complex entries for which $\begin{bmatrix} 0 & 1 \\ -1 & 0 \end{bmatrix} = P^{-1}\begin{bmatrix} i & 0 \\ 0 & -i \end{bmatrix}P$, see Exercise 9 (or try to guess one).

Similarly, the characteristic polynomial of $A = \begin{bmatrix} 1 & 1 \\ 1 & 2 \end{bmatrix}$ is $\begin{vmatrix} x-1 & -1 \\ -1 & x-2 \end{vmatrix} = x^2 - 3x + 1$. Since this polynomial has no rational roots (because it is monic, a rational root would have to be an integer—see Exercise 14—and it is easy to see that no integer is a root), it is irreducible, so $x^2 - 3x + 1$ is the sole elementary divisor of A and the rational canonical form of A is $\begin{bmatrix} 0 & -1 \\ 1 & 3 \end{bmatrix}$. If the "number" $\sqrt{5}$ is allowed, x^2-3x+1 has the factorization $(x-(3+\sqrt{5})/2)(x-(3-\sqrt{5})/2)$ and A is similar to the diagonal matrix $\begin{bmatrix} (3+\sqrt{5})/2 & 0 \\ 0 & (3-\sqrt{5})/2 \end{bmatrix}$ (see Exercise 10).

The theorem of Section 4 implies that there is a 3×3 matrix of rational numbers A such that $xI - A$ is equivalent to the diagonal matrix whose entries are $1, 1, (x-2)^3$. In fact, the construction used to prove that theorem gives the matrix $\begin{bmatrix} 0 & 0 & 8 \\ 1 & 0 & -12 \\ 0 & 1 & 6 \end{bmatrix}$ with the required property. The sole elementary divisor of this matrix is $(x-2)^3$, so its rational canonical form is $\begin{bmatrix} 2 & 0 & 0 \\ 1 & 2 & 0 \\ 0 & 1 & 2 \end{bmatrix}$ (see Proposition 2 of Section 5).

Similarly, the construction of Section 4 shows that

$$A = \begin{bmatrix} 0 & 0 & 0 & 0 & 0 \\ 1 & 0 & 0 & 0 & -1 \\ 0 & 1 & 0 & 0 & -3 \\ 0 & 0 & 1 & 0 & -4 \\ 0 & 0 & 0 & 1 & -3 \end{bmatrix}$$

has the property that $xI - A$ is equivalent to the diagonal matrix with entries $1, 1, 1, 1, x^5 + 3x^4 + 4x^3 + 3x^2 + x$. The factorization of this polynomial into irreducible factors is $x^5 + 3x^4 + 4x^3 + 3x^2 + x = (x^2 + x + 1)(x+1)^2 x$, so the elementary divisors of this matrix are $x^2 + x + 1$, $(x+1)^2$, x, and the rational canonical form is

$$\begin{bmatrix} 0 & -1 & 0 & 0 & 0 \\ 1 & -1 & 0 & 0 & 0 \\ 0 & 0 & -1 & 0 & 0 \\ 0 & 0 & 1 & -1 & 0 \\ 0 & 0 & 0 & 0 & 0 \end{bmatrix}$$

(or a variation of this matrix obtained using a different order for the three diagonal blocks $\begin{bmatrix} 0 & -1 \\ 1 & -1 \end{bmatrix}, \begin{bmatrix} -1 & 0 \\ 1 & -1 \end{bmatrix}, [\,0\,]$).

The matrices

$$\begin{bmatrix} 0 & -1 & 0 & 0 \\ 1 & 0 & 0 & 0 \\ 0 & 0 & 0 & -1 \\ 0 & 0 & 1 & 0 \end{bmatrix} \quad \text{and} \quad \begin{bmatrix} 0 & -1 & 0 & 0 \\ 1 & 0 & 0 & 0 \\ 1 & 0 & 0 & -1 \\ 0 & 1 & 1 & 0 \end{bmatrix}$$

are in rational canonical form. The first has elementary divisors $x^2 + 1, x^2 + 1$, while the second has the sole elementary divisor $(x^2 + 1)^2$. Therefore, the two matrices have the same characteristic polynomial $(x^2 + 1)^2$, but they are not similar. If complex numbers are allowed, the elementary divisors are $x - i$, $x - i, x + i, x + i$ in the case of the first matrix and $(x - i)^2$, $(x + i)^2$ in the case of the second, and the Jordan canonical forms of these matrices are

$$\begin{bmatrix} i & 0 & 0 & 0 \\ 0 & i & 0 & 0 \\ 0 & 0 & -i & 0 \\ 0 & 0 & 0 & -i \end{bmatrix} \quad \text{and} \quad \begin{bmatrix} i & 0 & 0 & 0 \\ 1 & i & 0 & 0 \\ 0 & 0 & -i & 0 \\ 0 & 0 & 1 & -i \end{bmatrix}$$

respectively.

Exercises

1 Find the rational canonical form C of the matrix $A = \begin{bmatrix} 0 & 1 \\ 1 & 0 \end{bmatrix}$ and find a matrix P with $A = P^{-1}CP$.

2 Prove that the relation of similarity (of square matrices of rational numbers) is reflexive, symmetric, and transitive.

3 The characteristic polynomial of a 3×3 matrix can be found without too much computation using the formula $|xI - A| = x^n - c_1x^{n-1} + c_2x^{n-2} - \cdots + (-1)^j c_j x^{n-j} + \cdots \pm c_n$, where c_j is the sum of the $j \times j$ **principal** minors of A, which is to say the $\binom{n}{j}$ minors of A formed by choosing j rows of A and *the same* j columns of A. Use this formula to find the characteristic polynomial of the 3×3 matrix $A = \begin{bmatrix} -3 & 1 & 1 \\ 1 & -1 & 1 \\ 1 & 1 & -3 \end{bmatrix}$. Conclude that this matrix is diagonalizable.

4 Use the formula of the preceding exercise to find the characteristic polynomial of

$$A = \begin{bmatrix} -28 & -5 & 15 \\ 13 & 1 & -6 \\ -48 & -9 & 26 \end{bmatrix}.$$

Verify the formula of the Cayley-Hamilton theorem for this A. What is the rational canonical form of this A?

5 Prove the formula $|xI - A| = x^n - c_1x^{n-1} + c_2x^{n-2} - \cdots + (-1)^j c_j x^{n-j} + \cdots \pm c_n$ of the preceding exercises.

6 Give two matrices in rational canonical form which have the same minimum polynomial but are not similar.

7 Find a matrix in rational canonical form similar to

$$A = \begin{bmatrix} 0 & 0 & 0 & 0 & -1 \\ 1 & 0 & 0 & 0 & 0 \\ 0 & 1 & 0 & 0 & 1 \\ 0 & 0 & 1 & 0 & 1 \\ 0 & 0 & 0 & 1 & 0 \end{bmatrix}.$$

8 What is wrong with the statement that $(x-2)^2$ is the minimum polynomial of the matrix

$$\begin{bmatrix} 1 & 0 & 0 \\ 0 & x-2 & 0 \\ 0 & 0 & (x-2)^2 \end{bmatrix}?$$

9 Use the method of the proof of Section 3 to find a matrix P for which

$$\begin{bmatrix} 0 & 1 \\ -1 & 0 \end{bmatrix} = P^{-1} \begin{bmatrix} i & 0 \\ 0 & -i \end{bmatrix} P.$$

10 Find a matrix P (whose entries necessarily involve $\sqrt{5}$) such that

$$\begin{bmatrix} 1 & 1 \\ 1 & 2 \end{bmatrix} = P^{-1} \begin{bmatrix} (3+\sqrt{5})/2 & 0 \\ 0 & (3-\sqrt{5})/2 \end{bmatrix} P.$$

11 Prove that a polynomial of degree n has at most n roots.

12 Let $f(x)$ be a polynomial with positive degree which is a multiple of a monic irreducible polynomial $g(x)$. Prove that $f(x)$ is a product of two relatively prime polynomials of positive degree unless it is a power of $g(x)$.

13 Prove that a matrix determines its elementary divisors. That is, if C_1 and C_2 are diagonal square matrices of polynomials in which each diagonal entry is monic and no diagonal entry can be written as a product of two relatively prime factors, and if C_1 and C_2 are equivalent, prove that C_2 can be obtained from C_1 merely by reordering the diagonal entries.

14 Show that a rational root of a monic polynomial with integer coefficients must be an integer.

15 Prove that if a product of two polynomials $f(x)g(x)$ is a multiple of $x - a$, where a is a rational number, then either $f(x)$ or $g(x)$ is a multiple of $x - a$. Conclude that if $f(x)$ divides a product of linear polynomials $(x - a_1)(x - a_2) \cdots (x - a_m)$ then $f(x)$ itself is a product of linear polynomials.

16 Prove the case $n = 1$, $A = 0$ of Proposition 2 of Section 5. (Find the minimum polynomial of $A^{(j)}$ in this case by direct means.)

17 Deduce the general case $n = 1$ of Proposition 2 of Section 5 from the case proved in the preceding exercise.

18 Prove Proposition 2 of Section 5. (Hint: Find the powers of $A^{(j)}$ and use them to determine the minimum polynomial of $A^{(j)}$.)

Chapter 10

The Spectral Theorem

1 Introduction

THE SPECTRAL THEOREM is not truly a theorem of linear algebra—or of algebra at all—because it involves in an essential way the use of *real numbers,* and real numbers involve *limits.* Nevertheless, it is covered in this last chapter because it is traditionally included in linear algebra courses, because it is one of the most beautiful and useful theorems in all of mathematics, and because the main content of the theorem can be formulated in a way that is purely algebraic (though not purely linear), until the limit notion enters at the last step (Section 7).

2 Orthogonal Partitions of Unity

A symmetric matrix that is equal to its own square is called an **orthogonal projection**. If A is a matrix and B is its mate, the matrix $P = AB$ is an orthogonal projection because $P = AB$ is symmetric and $P^2 = ABAB = AB = P$ by the definition of a mate. In this way, orthogonal projections already appeared in Chapter 7, even though they were not given their name there.

Two $n \times n$ orthogonal projections P and Q are said to be **orthogonal** to each other if $PQ = 0$. To say that $PQ = 0$ is the same as to say that $QP = 0$ because $QP = Q^T P^T = (PQ)^T$. The reason P and Q are called "orthogonal" when this condition is fulfilled is that if U is any $n \times 1$ matrix then the equality $\|(P+Q)U\|^2 = \big((P+Q)U\big)^T\big((P+Q)U\big) = U^T(P+Q)^T(P+Q)U = U^T(P+Q)(P+Q)U = U^T(P^2+PQ+QP+Q^2)U = U^T P^2 U + U^T Q^2 U = \|PU\|^2 + \|QU\|^2$ holds; geometrically, this equality means that PU and QU are the sides of a right triangle (the Pythagorean theorem), that is, that PU and QU are "orthogonal" for any U.

An **orthogonal partition of unity** is a set of nonzero orthogonal projections that are mutually orthogonal and add up to the identity matrix. That is, an orthogonal partition of unity is a set of $n \times n$ matrices $P_1, P_2, \ldots, P_k$ for which $I_n = P_1 + P_2 + \cdots + P_k$, $P_i \neq 0$, $P_i^T = P_i$, $P_i^2 = P_i$, and $P_i P_j = 0$ whenever $i \neq j$.

The matrix I_n by itself is an orthogonal partition of unity. Another very simple partition of unity consists of the n distinct $n \times n$ matrices with a single 1 on the diagonal and all other entries equal to 0. A less trivial example of an orthogonal partition of unity can be constructed by taking A to be a nonzero $n \times 1$ matrix ($n > 1$), letting B be the mate of A, and setting $P_1 = AB$, $P_2 = I - P_1$. It was noted above that a matrix P_1 obtained in this way is an orthogonal projection. Since neither A nor B is zero, $P_1 \neq 0$ because the entries of P_1 are the n^2 products of an entry of A and an entry of B. Clearly, P_2 is symmetric; it is an orthogonal projection because $P_2^2 = (I - P_1)(I - P_1) = I - P_1 - P_1(I - P_1) = I - P_1 - P_1 + P_1^2 = I - P_1 = P_2$. It is not zero because $P_2 = 0$ would imply $AB = I$, which is impossible because A has no right inverse. Moreover, P_1 and P_2 are orthogonal to each other because $P_1 P_2 = P_1(I - P_1) = P_1 - P_1 = 0$. Thus, $I = P_1 + P_2$ is an orthogonal partition of unity. This method of constructing orthogonal partitions of unity is the special case $P_1 = I_n$ of the construction in the proof of the following theorem.

Theorem. *In any orthogonal partition of unity $I_n = P_1 + P_2 + \cdots + P_k$, the ranks of the projections satisfy* $\text{rank}(P_1) + \text{rank}(P_2) + \cdots + \text{rank}(P_k) = n$. *Suppose the rank of some P_i is greater than* 1, *and suppose that $P_1, P_2, \ldots, P_k$ are reordered to make* $\text{rank}(P_1) > 1$. *Then there exist orthogonal projections Q_1 and Q_2 such that $P_1 = Q_1 + Q_2$ and such that $I_n = Q_1 + Q_2 + P_2 + \cdots + P_k$ is an orthogonal partition of unity.*

Proof. The **trace** of a square matrix P, denoted $\text{tr}(P)$, is the sum of its diagonal entries. *The trace of an orthogonal projection is equal to its rank.* As was shown in Chapter 7 (Section 3), if P is an $n \times n$ matrix whose rank is r, then P can be written as a product $P = TUV$ in which T is an $n \times r$ matrix with a left inverse, U is an invertible $r \times r$ matrix, and V is an $r \times n$ matrix with a right inverse; then the mate of P is $V^M U^M T^M$. If P is an orthogonal projection, it is its own mate (because PP is symmetric and $PPP = PP = P$) and $P = P^2 = PP^M = TUVV^MU^MT^M = TT^M$ (because VV^M and UU^M are both I_r), so the trace of P is the trace of TT^M. But the sum of the diagonal entries of TT^M is the sum over all $i = 1, 2, \ldots, n$ of the sums $t_{i1}u_{1i} + t_{i2}u_{2i} + \cdots + t_{ir}u_{ri}$ where t_{ij} is the entry in row i of column j of T and u_{ji} is the entry in row j of column i of T^M. In short, $\text{tr}(P)$ is the sum of the rn numbers $t_{ij}u_{ji}$. But the sum of these rn numbers is also equal to the trace of T^MT (because the diagonal entries of T^MT are the sums $u_{j1}t_{1j} + u_{j2}t_{2j} + \cdots + u_{jn}t_{nj}$ for $j = 1, 2, \ldots, r$), that is, $\text{tr}(P) = \text{tr}(T^MT) = \text{tr}(I_r) = r$, as was to be shown.

Since the trace of the sum of a set of matrices is the sum of their traces (because the diagonal elements of the sum are the sums of the diagonal elements), the identity $\text{rank}(P_1)+\text{rank}(P_2)+\cdots+\text{rank}(P_k) = \text{tr}(P_1)+\text{tr}(P_2)+\cdots+\text{tr}(P_k) = \text{tr}(I_n) = n$ follows.

Suppose now that $\text{tr}(P_1) > 1$. Let A be an $n \times 1$ matrix which is a nonzero column of P_1, let $Q_1 = AA^M$ be the product of A and its mate, and let $Q_2 = P_1 - Q_1$. Then Q_1 is an orthogonal projection. The explicit formula $Q_1 = Aa^{-1}A^T$, where a is the 1×1 matrix A^TA, shows that the trace of Q_1 is the trace of $a^{-1}AA^T$ which is $a^{-1}A^TA = 1$. Therefore the rank of Q_1 is 1, which implies that $Q_1 \neq 0$ and $Q_2 \neq 0$ (because $Q_2 = 0$ would imply $Q_1 = P_1$ and the rank of P_1 is greater than 1). Moreover, $P_1P_1 = P_1$ implies, since A is a column of P_1, that $P_1A = A$, and consequently that $P_1Q_1 = P_1AA^M = AA^M = Q_1$. The transpose of this equation is $Q_1P_1 = Q_1$. Therefore $Q_2^2 = (P_1 - Q_1)(P_1 - Q_1) = P_1(P_1 - Q_1) - Q_1(P_1 - Q_1) = P_1 - Q_1 - Q_1 + Q_1 = Q_2$. Since Q_2 is a difference of symmetric matrices, it a symmetric matrix, so Q_2 is a nonzero orthogonal projection. The orthogonal projections Q_1 and Q_2 are orthogonal because $Q_1Q_2 = Q_1(P_1 - Q_1) = Q_1 - Q_1 = 0$. Because A is a column of P_1, $P_iP_1 = 0$ implies $P_iA = 0$ and therefore implies $P_iQ_1 = 0$ and $P_iQ_2 = P_i(P_1 - Q_1) = 0 - 0 = 0$ for $i > 1$. Therefore, $Q_1, Q_2, P_2, P_3, \ldots, P_k$ are mutually orthogonal, nonzero, and their sum is I_n, as was to be shown.

The construction used in the proof gives an easy method of constructing orthogonal partitions of unity.

3 *Spectral Representations*

If $P_1 + P_2 + \cdots + P_k = I_n$ is an orthogonal partition of unity and $a_1, a_2, \ldots, a_k$ are rational numbers, then $a_1P_1+a_2P_2+\cdots+a_kP_k$ is a symmetric matrix. (Each term a_iP_i of the sum is a symmetric matrix.) A representation of a symmetric matrix S in this form $S = a_1P_1 + a_2P_2 + \cdots + a_kP_k$ is called* a **spectral representation** of S.

Theorem. *Let $S = a_1P_1 + a_2P_2 + \cdots + a_kP_k$ be a spectral representation of a symmetric matrix S. Then the minimum polynomial of S is the product of the* distinct *polynomials in the list $x - a_1, x - a_2, \ldots, x - a_k$.*

Proof. If two of the numbers $a_1, a_2, \ldots, a_k$ are equal, say $a_1 = a_2$, then $a_1(P_1 + P_2)+a_3P_3+\cdots+a_kP_k$ is another spectral representation of the same symmetric matrix in which a_1 and a_2 have been consolidated. Since this process can be repeated as long as any two of the as are equal, there is no loss of generality in

*The terminology derives from an analogy to atomic spectra long forgotten by most mathematicians.

assuming that the given spectral representation $S = a_1 P_1 + a_2 P_2 + \cdots + a_k P_k$ is one in which the numbers $a_1, a_2, \ldots, a_k$ are distinct.

Expansion of $S^2 = (a_1 P_1 + a_2 P_2 + \cdots + a_k P_k)(a_1 P_1 + a_2 P_2 + \cdots + a_k P_k)$ gives S^2 as a sum of k^2 terms $a_i a_j P_i P_j$ in which all but the k terms $a_i^2 P_i^2 = a_i^2 P_i$ are zero. Therefore, $S^2 = a_1^2 P_1 + a_2^2 P_2 + \cdots + a_k^2 P_k$. A similar argument shows that $S^3 = (a_1^2 P_1 + a_2^2 P_2 + \cdots + a_k^2 P_k)(a_1 P_1 + a_2 P_2 + \cdots + a_k P_k) = a_1^3 P_1 + a_2^3 P_2 + \cdots + a_k^3 P_k$ and, more generally, $S^l = a_1^l P_1 + a_2^l P_2 + \cdots + a_k^l P_k$ for any positive integer l. Therefore, for any polynomial $h(x) = c_m x^m + c_{m-1} x^{m-1} + \cdots + c_0$, the formula

$$\begin{aligned}
h(S) &= c_m S^m + c_{m-1} S^{m-1} + \cdots + c_0 I \\
&= c_m (a_1^m P_1 + \cdots + a_k^m P_k) + c_{m-1}(a_1^{m-1} P_1 + \cdots + a_k^{m-1} P_k) + \cdots \\
&\qquad \cdots + c_0 (P_1 + \cdots + P_k) \\
&= (c_m a_1^m + c_{m-1} a_1^{m-1} + \cdots + c_0) P_1 + \cdots \\
&\qquad \cdots + (c_m a_k^m + c_{m-1} a_k^{m-1} + \cdots + c_0) P_k \\
&= h(a_1) P_1 + h(a_2) P_2 + \cdots + h(a_k) P_k
\end{aligned}$$

holds.

Let $f(x) = (x - a_1)(x - a_2) \cdots (x - a_k)$ and let $g(x)$ be the minimum polynomial of S. The formula just proved shows that $f(S) = 0$ and therefore that $f(x)$ is a multiple of $g(x)$. On the other hand, the formula also shows that $g(a_i) P_i = (g(a_1) P_1 + \cdots + g(a_k) P_k) P_i = g(S) P_i = 0$; since $P_i \neq 0$, it follows that $g(a_i) = 0$ for $i = 1, 2, \ldots, k$. Because a polynomial of degree m has at most m distinct roots,* it follows that the degree of $g(x)$ is at least k. Thus, $f(x)$, which is monic of degree k, is a multiple of $g(x)$, which is monic of degree at least k, which implies $f(x) = g(x)$, as was to be shown.

Thus, if a symmetric matrix S has a spectral representation, its minimum polynomial is a product of distinct factors $x - a_i$ of the first degree. As was seen in Section 7 of Chapter 9, the minimum polynomial of S has this form if and only if S is *diagonalizable* (similar to a diagonal matrix). The theorem implies, then, that *a symmetric matrix which has a spectral representation must be diagonalizable.*

4 Symmetric Matrices with Spectral Representations

The necessary condition that was found in the last section for a symmetric matrix S to have a spectral representation is in fact sufficient, that is:

*See Exercise 11 of Chapter 9.

Theorem. *If a symmetric matrix is diagonalizable, it has a spectral representation.*

Proof. Let S be a diagonalizable symmetric matrix. Then the minimum polynomial of S has the form $f(x) = (x - a_1)(x - a_2)\cdots(x - a_k)$ where the numbers $a_1, a_2, \ldots, a_k$ are distinct. For each $i = 1, 2, \ldots, k$, let $g_i(x) = f(x)/(x - a_i)$; that is, let $g_i(x)$ be the product of the linear factors of $f(x)$ with the ith factor $x - a_i$ left out. Let $G_i = g_i(S)$, let H_i be the mate of G_i, and let $P_i = G_i H_i$.

Then each P_i is an orthogonal projection. Since $g_i(x)$ is not a multiple of the minimum polynomial $f(x)$ of S (a_i is a root of $f(x)$ but not of $g_i(x)$) $G_i \neq 0$, so the formula $G_i = G_i H_i G_i = P_i G_i$ shows that $P_i \neq 0$.

For $i \neq j$, $G_i G_j$ is a multiple of $f(S)$ (because G_j contains the one factor $S - a_i I$ of $f(S)$ that is missing from G_i); since $f(S) = 0$, it follows that $G_i G_j = 0$. Therefore, $G_i P_j = G_i G_j H_j = 0$. Since both G_i and P_j are symmetric (G_i is a polynomial in the symmetric matrix S), transposition of $G_i P_j = 0$ gives $P_j G_i = 0$. Therefore, $P_j P_i = P_j G_i H_i = 0$, that is, P_i and P_j are orthogonal when $i \neq j$.

That $I = P_1 + P_2 + \cdots + P_k$ can be proved as follows. Let $d(x)$ be the greatest common divisor of $g_1(x), g_2(x), \ldots, g_k(x)$. For each i, $g_i(a_i)$ is the product of the $k - 1$ nonzero numbers $a_i - a_j$ in which $j \neq i$. Therefore, $g_i(a_i) \neq 0$. Since $g_i(x) = q_i(x)d(x)$ for some polynomial $q_i(x)$, it follows that $d(a_i) \neq 0$ for $i = 1, 2, \ldots, k$. On the other hand, $g_1(a_j) = 0$ whenever $j \neq 1$, so the equation $g_1(a_j) = q_1(a_j)d(a_j)$ combines with $d(a_j) \neq 0$ to give $q_1(a_j) = 0$ whenever $j \neq 1$. Therefore, $q_1(x)$ has $k - 1$ distinct roots $a_2, a_3, \ldots, a_k$, which implies that the degree of $q_1(x)$ is at least $k - 1 = \deg g_1$. Thus, $g_1(x) = q_1(x)d(x)$ implies that the degree of $q_1(x)$ is exactly $k - 1$, and $d(x)$ is a nonzero constant; since $d(x)$ is monic, it is 1. Thus, since $d(x)$ is a sum of multiples of the $g_i(x)$, there exist polynomials $n_1(x), n_2(x), \ldots, n_k(x)$ such that $1 = g_1(x)n_1(x) + g_2(x)n_2(x) + \cdots + g_k(x)n_k(x)$. Substitution of S for x in the polynomial $1 - g_1(x)n_1(x) - g_2(x)n_2(x) - \cdots - g_k(x)n_k(x) = 0$ gives $I - \sum_{i=1}^{k} G_i N_i = 0$, that is, gives $I = \sum G_i N_i = \sum G_i H_i G_i N_i = \sum P_i G_i N_i$, where $N_i = n_i(S)$ and the sum is a sum of the k terms $P_i G_i N_i$ for $i = 1, 2, \ldots, k$. Multiplication of this equation on the left by P_j gives $P_j = P_j P_j G_j N_j = P_j G_j N_j$ for each j ($P_j P_i = 0$ when $i \neq j$). Substitution of these k equations in $I = \sum P_i G_i N_i$ gives $I = \sum P_i$, as desired. Thus, $I = P_1 + P_2 + \cdots + P_k$ is an orthogonal partition of unity.

Substitution of S in $(x - a_i)g_i(x) = f(x)$ gives $(S - a_i I)G_i = 0$. Therefore, $(S - a_i I)P_i = 0$, that is, $SP_i = a_i P_i$, and multiplication of $I = P_1 + P_2 + \cdots + P_k$ by S gives the spectral representation $S = a_1 P_1 + a_2 P_2 + \cdots + a_k P_k$.

It is in fact easy to see that there is *only one* spectral representation of S in which the as are distinct, provided one disregards the order of the terms in the sum $S = a_1P_1 + a_2P_2 + \cdots + a_kP_k$. In the first place, S determines its minimum polynomial, so by the theorem of the preceding section S determines the distinct coefficients $a_1, a_2, \ldots, a_k$ in any spectral representation. Thus, any spectral representation of S in which the as are distinct is, up to the order of the terms, of the form $S = a_1P_1' + a_2P_2' + \cdots + a_kP_k'$ where $I_n = P_1' + P_2' + \cdots + P_k'$ is an orthogonal partition of unity. Such a spectral representation implies, as above, that if $h(x)$ is any polynomial then $h(S) = h(a_1)P_1' + h(a_2)P_2' + \cdots + h(a_k)P_k'$. Thus, the matrix G_i constructed in the proof just given is equal to $g_i(a_1)P_1' + g_i(a_2)P_2' + \cdots + g_i(a_k)P_k' = g_i(a_i)P_i'$ (because $g_i(a_j) = 0$ when $i \neq j$). It follows directly from the definition of a mate that $H_i = g_i(a_i)^{-1}P_i'$, so $P_i = G_iH_i = P_i'$, as was to be shown.

5 *Sign Changes in Polynomials*

The symmetric matrix $\begin{bmatrix} 1 & 1 \\ 1 & 2 \end{bmatrix}$ with minimum polynomial $x^2 - 3x + 1$ (see the Examples section of Chapter 9) does not have a spectral representation because the roots

$$a_1, a_2 = \frac{3 \pm \sqrt{5}}{2}$$

of its minimum polynomial are irrational numbers. However, if $\sqrt{5}$ is admitted as a "number," this matrix is diagonalizable and the construction in the last section gives a spectral representation of it. (See the Examples section.) The fact that this polynomial has real—as opposed to imaginary—roots follows from the simple observation that $f(x) = x^2 - 3x + 1$ is positive when $x = 0$, negative when $x = 2$, and positive again when $x = 4$. It will be shown in the next section that the minimum polynomial of any symmetric matrix behaves in a similar way. The present section is devoted to a classical theorem about sign changes in polynomials.

Sturm's Theorem. *Let $s_0(x), s_1(x), \ldots, s_m(x)$ be a sequence of polynomials with rational coefficients for which*

(i) s_i is a polynomial of degree i with positive leading coefficient, and

(ii) $s_{i+1}(x) + s_{i-1}(x)$ is a multiple of $s_i(x)$ for $i = 1, 2, \ldots, m-1$.

Then $s_m(x)$ changes sign m times. That is, there is a sequence of $m+1$ rational numbers $r_0 < r_1 < \cdots < r_m$ such that the $m+1$ rational numbers $s_m(r_0)$, $s_m(r_1)$, $\ldots$, $s_m(r_m)$ are all nonzero and alternate in sign.

Each s_i satisfies the same conditions as s_m, so the theorem implies that s_i changes sign i times.

Proof. Since a polynomial of degree i has at most i roots, each of the numbers $s_0(x), s_1(x), \ldots, s_m(x)$ has a well-defined *sign* for all rational values of x except at most $0+1+2+\cdots+m$. For values of x other than these few, let $J(x)$ be defined to be the number of sign changes in the sequence of rational numbers $s_0(x), s_1(x), s_2(x), \ldots, s_m(x)$.

When x is very large, the leading term of $s_i(x)$ dominates all the others, and $s_i(x)$ is positive because its leading term is positive. Therefore $J(x)=0$ when x is sufficiently large. Similarly, for large x, the leading term of $s_i(-x)$ dominates all the others, and the sign of $s_i(-x)$ is the same as the sign of the leading term, which is $(-1)^i$. Therefore, $J(-x)=m$ for all sufficiently large numbers x. Let a and b be numbers such that $J(a)=m$ and $J(b)=0$.

With a and b chosen in this way, let the interval $a \le x \le b$ be partitioned into a very large number of very small subintervals by the insertion of a large number of intermediate points, at all of which $J(x)$ is defined. Since $J(x)$ changes only when one of the polynomials $s_i(x)$ changes sign, $J(x)$ does not change on most of the small subintervals. It will be shown that when the subintervals are sufficiently small the change in $J(x)$ on any one subinterval is at most 1, and it is zero unless the sign of $s_m(x)$ changes on that subinterval.

Loosely speaking, this is true for the following reasons. Condition (ii) implies that if both $s_i(x)$ and $s_{i+1}(x)$ are zero for a particular x then $s_{i-1}(x), s_{i-2}(x), \ldots,$ and $s_0(x)$ are also zero for that x, which is impossible because $s_0(x)$ is a positive constant. Therefore, a very short interval that contains a zero of $s_i(x)$ will not contain a zero of $s_{i+1}(x)$ or of $s_{i-1}(x)$. Thus, if some $s_i(x)$ other than $s_m(x)$ changes sign on such an interval—which implies it has a zero in the interval—both $s_{i+1}(x)$ and $s_{i-1}(x)$ are nonzero throughout the interval and, by condition (ii), they have *opposite signs*. Thus, the sequence $s_{i-1}(x), s_i(x), s_{i+1}(x)$ changes sign exactly once, regardless of the sign of $s_i(x)$. Therefore, on a short interval, $J(x)$ can change only if $s_m(x)$ changes its sign, and, when this happens, $J(x)$ changes by just one.

Readers experienced in calculus will be able to make a rigorous proof based on these ideas. A proof without calculus can be given as follows. Since, for each n, $x^n - y^n = (x-y)(x^{n-1}+x^{n-2}y+x^{n-3}y^2+\cdots+y^{n-1})$, each difference $s_i(x)-s_i(y)$ can be written in the form $(x-y)G_i(x,y)$, where G_i is a polynomial in two variables with rational coefficients. For each $i = 1, 2, \ldots, m$ there is a positive number B_i such that $|G_i(x,y)| < B_i$ for all pairs of numbers (x,y) with $a \le x \le b$ and $a \le y \le b$. If B is the largest of the numbers B_i, then $|s_i(x)-s_i(y)| < |x-y|B$ for all $i = 1, 2, \ldots, m$ and for all x and y between a and b. By (ii), a common divisor of $s_{i+1}(x)$ and $s_i(x)$ divides $s_{i-1}(x), s_{i-2}(x), \ldots, s_0(x)$, so it must have degree zero. Therefore, 1 is the greatest common divisor of $s_{i+1}(x)$ and $s_i(x)$, so one can find, for each i, polynomials $g_i(x)$ and

$h_i(x)$ such that $g_i(x)s_{i+1}(x) + h_i(x)s_i(x) = 1$. Let C be a positive number greater than all values of all of the polynomials $g_i(x)$ and $h_i(x)$ for $a \le x \le b$. Finally, let $q_i(x)$ be the polynomial defined by $s_{i+1}(x) + s_{i-1}(x) = q_i(x)s_i(x)$, and let D be a number greater than 2 and greater than all values of $q_i(x)$ for $a \le x \le b$. If l is a positive number less than $\frac{1}{BCD}$, if c and d are rational numbers satisfying $|c - d| \le l$ for which $J(c)$ and $J(d)$ are both defined, and if $J(c) \ne J(d)$, then $|J(c) - J(d)| = 1$ and $s_m(c)$ and $s_m(d)$ have opposite signs, as can be seen as follows.

Suppose $s_i(c)$ and $s_i(d)$ have opposite signs. If x is in the interval $c \le x \le d$ and $s_i(x) \ne 0$, then either $s_i(c)$ or $s_i(d)$ has the sign opposite to that of $s_i(x)$, so either $|s_i(x)| \le |s_i(x) - s_i(c)|$ or $|s_i(x)| \le |s_i(x) - s_i(d)|$; since both $|x - c| \le l$ and $|x - d| \le l$, it follows in either case that $|s_i(x)| < lB \le \frac{1}{CD}$, that is, all values of s_i in the interval $c \le x \le d$ are less than $\frac{1}{CD}$ in absolute value. Therefore, $1 = |g_{i-1}(x)s_i(x) + h_{i-1}(x)s_{i-1}(x)| \le C\frac{1}{CD} + C|s_{i-1}(x)|$, so $|s_{i-1}(x)| \ge C^{-1}(1 - \frac{1}{D}) = \frac{D-1}{CD} > \frac{2-1}{CD} = \frac{1}{CD}$ for all x in the interval. In particular, s_{i-1} is never zero in the interval $c \le x \le d$. (Note that $i > 0$ because s_0 is constant.) Moreover, $s_{i-1}(c)$ and $s_{i-1}(d)$ must have the same sign (opposite signs would imply $|s_{i-1}(x)| \le \frac{1}{CD}$, as was just shown). When $i < m$, the same method shows that s_{i+1} is never zero in the interval and that $s_{i+1}(c)$ and $s_{i+1}(d)$ have the same sign.

Moreover, if $i < m$ and if $s_i(c)$ and $s_i(d)$ have opposite signs, then $s_{i-1}(x)$ and $s_{i+1}(x)$ have opposite signs throughout the interval $c \le x \le d$, because if they had the same sign for some x then $2\frac{D-1}{CD} \le |s_{i+1}(x)| + |s_{i-1}(x)| = |s_{i+1}(x) + s_{i-1}(x)| = |q_i(x)s_i(x)| \le D\frac{1}{CD}$ and therefore $2D - 2 \le D$, contrary to the assumption that $D > 2$. Therefore, the sequence of three rational numbers $s_{i-1}(x)$, $s_i(x)$, $s_{i+1}(x)$ contains exactly *one* change of sign for any x in the interval $c \le x \le d$ for which $s_i(x) \ne 0$. In particular, the sequences $s_{i-1}(c)$, $s_i(c)$, $s_{i+1}(c)$ and $s_{i-1}(d)$, $s_i(d)$, $s_{i+1}(d)$ both contain exactly one sign change.

If $s_m(c)$ and $s_m(d)$ have the same sign, then $J(c) = J(d)$ because J can only change if the sign of at least one s_i changes, but s_m does not change by assumption, and, as the preceding paragraph shows, a change in the sign of s_i for $i < m$ does not change J. Thus, J does not change unless the sign of s_m changes. When the sign of s_m does change, the sign of s_{m-1} does not change, and the number of sign changes in the sequences $s_0(c), s_1(c), \ldots, s_{m-1}(c)$ and $s_0(d)$, $s_1(d), \ldots, s_{m-1}(d)$ are the same. Therefore, $J(c)$ and $J(d)$ differ by exactly one.

Let the interval $a \le x \le b$ be partitioned into subintervals, all of length less than l, whose endpoints are points where $J(x)$ is defined. Since $J(x)$ undergoes a change of m on the whole interval $a \le x \le b$, $J(x)$ must change on at least m

of the subintervals, and therefore $s_m(x)$ must reverse its sign on at least m of the subintervals, which proves the theorem.

A polynomial of degree m changes sign at most m times (see Exercise 10). Therefore, $J(x)$ can change on at most m subintervals; since $J(x)$ must decrease from m to 0, each change must *decrease* $J(x)$. Thus, each change in the sign of s_m must be a change from disagreement to agreement with the sign of s_{m-1}. This observation implies that s_{m-1} must change sign between successive sign changes of s_m, and, more generally, that s_{i-1} must change sign between successive sign changes of s_i.

Note that the method of the proof can be used to *find* sign changes in $s_m(x)$. The value of $J(x)$—which is easy to find—is the number of sign changes of $s_m(x)$ between x and b. Starting with the interval $a \leq x \leq b$ and successively bisecting it, one can use this fact to find intervals on which $s_m(x)$ has only one sign change. In this way, numbers $r_0 < r_1 < \cdots < r_m$ with the desired property can be found very quickly.

6 *The Algebraic Theorem*

Theorem. *The minimum polynomial of a symmetric matrix with rational entries changes sign a number of times equal to its degree.*

Proof. Let S be an $n \times n$ symmetric matrix with rational entries, let $f(x)$ be its minimum polynomial, and let $m = \deg f$. The method of the proof will be to find a sequence of polynomials $s_0(x), s_1(x), \ldots, s_m(x)$ satisfying conditions (i) and (ii) of Sturm's theorem in the preceding section and satisfying $s_m(x) = f(x)$. That $f(x)$ has the required number of sign changes will then follow from Sturm's theorem.

Once $s_m(x)$ and $s_{m-1}(x)$ are known, all the others are determined by condition (ii). Indeed, any two polynomials $f(x)$ and $g(x)$, with $g(x) \neq 0$, determine polynomials $q(x)$ and $r(x)$ such that $f(x) = q(x)g(x) + r(x)$ and $\deg r < \deg g$ (Exercise 9). Application of this observation to the equation $s_{i+1}(x) = q_i(x)s_i(x) - s_{i-1}(x)$ shows, because $\deg s_{i-1} < \deg s_i$, that $s_{i+1}(x)$ and $s_i(x)$ determine both $q_i(x)$ and $s_{i-1}(x)$. Therefore, $s_m(x)$ and $s_{m-1}(x)$ determine all the others in succession. What is needed, then, is a polynomial $s_{m-1}(x)$ with the property that the sequence of polynomials $s_{m-2}(x), s_{m-3}(x), \ldots$ determined by it and by $s_m(x) = f(x)$ has property (i).

If A and B are square matrices of the same size and a and b are numbers, then clearly* $\mathrm{tr}(aA + bB) = a\,\mathrm{tr}(A) + b\,\mathrm{tr}(B)$. For any symmetric matrix T, $\mathrm{tr}(T^2) = \|T\|^2$ because,† by the symmetry of T, the ith diagonal entry of T^2 is equal to the sum of the squares of the entries in the ith row (column) of T. In particular, $\mathrm{tr}(T^2) > 0$ for every symmetric matrix T other than $T = 0$.

Let B be the $m \times m$ matrix whose entry in the ith row and the jth column is $\mathrm{tr}(S^{2m-i-j})$. For any pair of polynomials $g(x) = g_1 x^{m-1} + g_2 x^{m-2} + \cdots + g_m$ and $h(x) = h_1 x^{m-1} + h_2 x^{m-2} + \cdots + h_m$ of degree less than m, the formula $\mathrm{tr}(g(S)h(S)) = G^T B H$ holds, where G and H are the $m \times 1$ matrices with entries $g_1, g_2, \ldots, g_m$ and $h_1, h_2, \ldots, h_m$, respectively. If B were not invertible, it would be a zero divisor, and there would be a nonzero H such that $BH = 0$. But this is impossible, because $H \neq 0$ implies $h(x) \neq 0$, which implies that $h(x)$ is not a multiple of $f(x)$ (its degree is less than $m = \deg f$) and therefore that $h(S) \neq 0$, contrary to $\mathrm{tr}(h(S)^2) = H^T B H = 0$. Therefore, B is invertible.

Let $H = B^{-1}E$ where E is the $m \times 1$ matrix whose entries are 1, 0, 0, ..., 0, and let $h(x) = h_1 x^{m-1} + h_2 x^{m-2} + \cdots + h_m$ be the polynomial whose coefficients are the entries of H. For this $h(x)$ and for any polynomial $g(x) = g_1 x^{m-1} + g_2 x^{m-2} + \cdots + g_m$ of degree less than m, the equation $\mathrm{tr}(g(S)h(S)) = G^T B H = G^T E = g_1$ holds.

With this choice of $h(x)$, let $d(x)$ be the greatest common divisor of $f(x)$ and $h(x)$. The degree of $d(x)$ must be 0—that is, $d(x)$ must be 1—for the following reason. If $d(x)$ had positive degree, it would be a common factor of $f(x)$ and $h(x)$, say $f(x) = b(x)d(x)$ and $h(x) = c(x)d(x)$, where $t = \deg d > 0$ and $\deg b = m - t < m$. Then $\mathrm{tr}(S^{t-1}b(S)h(S))$ on the one hand would be 1 (because it is the coefficient of x^{m-1} in $x^{t-1}b(x)$, which is the leading coefficient of $b(x)$) and on the other hand would be 0 (because it is $\mathrm{tr}(S^{t-1}b(S)c(S)d(S)) = \mathrm{tr}(S^{t-1}f(S)c(S)) = \mathrm{tr}(0)$). Thus, there are polynomials $u(x)$ and $v(x)$ such that $1 = u(x)f(x) + v(x)h(x)$. Let $s_{m-1}(x)$ be the remainder when $v(x)$ is divided by $f(x)$, that is, $v(x) = q(x)f(x) + s_{m-1}(x)$, where $q(x)$ is a polynomial and $s_{m-1}(x)$ is a polynomial of degree less than m.

It will be shown that this $s_{m-1}(x)$ has the required properties. Let $s_{m-2}(x)$, $s_{m-3}(x)$, ... be defined by the conditions that $s_{i+1}(x) + s_{i-1}(x) = q_i(x)s_i(x)$ and $\deg s_{i-1} < \deg s_i$. Since $\deg s_i$ is decreasing, a stage is necessarily reached at which $s_i(x) = 0$. Further terms are then undefined, because division by 0 is meaningless. The terms in the sequence after $s_i(x) = 0$ will be defined to be zero. Let c_i be the coefficient of x^i in $s_i(x)$. Since $\deg s_i \leq i$, what is to be shown is simply that $c_i > 0$ for $i = m-1, m-2, \ldots, 1, 0$.

---

*Recall that the trace of a square matrix A, denoted $\mathrm{tr}(A)$, is the sum of its diagonal entries.
†Recall that $\|A\|^2$ denotes the sum of the squares of the entries of a matrix A.

Since $1 = f(x)u(x) + h(x)(q(x)f(x) + s_{m-1}(x))$, $h(x)s_{m-1}(x) - 1$ is a multiple of $f(x)$, from which it follows that $h(S)s_{m-1}(S) - I_n = 0$, that is, the matrix $s_{m-1}(S)$ is the inverse of $h(S)$. Thus $c_{m-1} = \text{tr}(s_{m-1}(S)h(S)) = \text{tr}(I_n) = n > 0$.

If $m = 1$ there is nothing more to prove. Otherwise,

$$\begin{aligned}
\text{tr}(s_{m-2}(S)^2h(S)^2) &= \text{tr}(s_{m-2}(S)\big(q_{m-1}(S)s_{m-1}(S) - s_m(S)\big)h(S)^2) \\
&= \text{tr}(s_{m-2}(S)q_{m-1}(S)s_{m-1}(S)h(S)^2) \\
&\quad - \text{tr}(s_{m-2}(S)f(S)h(S)^2) \\
&= \text{tr}(s_{m-2}(S)q_{m-1}(S)h(S)) \\
&= \text{coefficient of } x^{m-1} \text{ in } s_{m-2}(x)q_{m-1}(x)
\end{aligned}$$

which shows that the coefficient of x^{m-1} in $s_{m-2}(x)q_{m-1}(x)$ is positive. Now $q_{m-1}(x)$, being the quotient when $f(x) = x^m + \cdots$ is divided by $s_{m-1}(x) = nx^{m-1} + \cdots$, is a polynomial of degree 1 with leading coefficient $1/n$. Therefore, the coefficient of x^{m-1} in $s_{m-2}(x)q_{m-1}(x)$ is c_{m-2}/n. Thus, $c_{m-2}/n > 0$, so $c_{m-2} > 0$.

If $m = 2$, there is nothing more to prove. The calculation just given can be generalized to show that for any polynomial $g(x)$ of degree less than $m-1$, $\text{tr}(g(S)s_{m-2}(S)h(S)^2) = \text{tr}(g(S)\big(q_{m-1}(S)s_{m-1}(S) - s_m(S)\big)h(S)^2) = \text{tr}(g(S)q_{m-1}(S)h(S))$ is the coefficient of x^{m-1} in $g(x)q_{m-1}(x)$, that is, $1/n$ times the coefficient of x^{m-2} in $g(x)$. Therefore, for any $g(x)$ of degree less than $m-2$,

$$\begin{aligned}
\text{tr}(g(S)s_{m-3}(S)h(S)^2) &= \text{tr}(g(S)\big(q_{m-2}(S)s_{m-2}(S) - s_{m-1}(S)\big)h(S)^2) \\
&= \text{tr}(g(S)q_{m-2}(S)s_{m-2}(S)h(S)^2) - \text{tr}(g(S)h(S)) \\
&= \text{tr}(g(S)q_{m-2}(S)s_{m-2}(S)h(S)^2) \\
&= \tfrac{1}{n}\big(\text{coefficient of } x^{m-2} \text{ in } g(x)q_{m-2}(x)\big) \\
&= \frac{\text{coefficient of } x^{m-3} \text{ in } g(x)}{c_{m-2}}
\end{aligned}$$

because $q_{m-2}(x)$, being the quotient when $s_{m-1}(x) = nx^{m-1} + \cdots$ is divided by $s_{m-2}(x) = c_{m-2}x^{m-2} + \cdots$, is a polynomial of degree 1 with leading coefficient n/c_{m-2}. In particular, $c_{m-3}/c_{m-2} = \text{tr}(s_{m-3}(S)^2h(S)^2) > 0$, which proves that $c_{m-3} > 0$.

If $m = 3$, there is nothing more to prove. Otherwise, the process can be continued. The crucial formula states that if $g(x) = b_{i-1}x^{i-1} + \cdots$ is any polynomial of degree less than i then $\text{tr}(g(S)s_{i-1}(S)h(S)^2) = b_{i-1}/c_i$. This was proved above in the cases $i = m-1$ and $m-2$. Suppose it is true for $i+1$ and suppose $c_{i+1} > 0$. Then $\text{tr}(s_i(S)^2h(S)^2) = c_i/c_{i+1}$ shows that $c_i > 0$. Moreover, since it follows from this that $q_i(x)$ has degree 1 and leading coefficient c_{i+1}/c_i

$$\begin{aligned}\text{tr}(g(S)s_{i-1}(S)h(S)^2) &= \text{tr}(g(S)(q_i(S)s_i(S) - s_{i+1}(S))h(S)^2)\\ &= \frac{b_{i-1}(c_{i+1}/c_i)}{c_{i+1}} - 0 = \frac{b_{i-1}}{c_i},\end{aligned}$$

so the formula is true for i. Therefore, $c_i > 0$ for $i = m-3, m-4, \ldots, 1, 0$, as was to be shown.

7 *Real Numbers*

The spectral theorem states that if the notion of "number" is widened to mean *real numbers,* then any symmetric matrix S has a spectral representation. This statement follows easily from what has been proved above, once the notion of "real number" is explained.

A book on linear algebra is not the place for a detailed explanation of real numbers. Suffice it to say that a real number is a number that can be expressed as the limit of a convergent sequence of rational numbers. Rational numbers are also real numbers, and the operations of arithmetic—addition, subtraction, multiplication, and division—can be applied to real as well as to rational numbers. In the language of modern algebra, the real numbers form a *field* containing the field of rational numbers. Moreover, they form an *ordered* field in that the relation $a > b$ has a natural meaning for real numbers a and b that extends its meaning for rational numbers. (Namely, if a_n is a sequence of rational numbers converging to the real number a and if b_n is a sequence of rational numbers converging to the real number b, then $a > b$ means there is a positive rational number ϵ such that $a_n > b_n + \epsilon$ for all sufficiently large n.) A final fact about real numbers of importance to the spectral theorem is that they share with the rational numbers the property that the sum of the squares of a finite set of real numbers is *positive* unless the numbers are all zero.

Real numbers do, however, have a crucial *intermediate value property* that the rational numbers lack: If $f(x)$ is a polynomial whose coefficients are real numbers, and if a and b are real numbers such that $f(a) < 0$ and $f(b) > 0$, then there is a real number c between a and b such that $f(c) = 0$. For example, the value of the polynomial $x^2 - 2$ is negative when $x = 1$ and positive when $x = 2$; there is a *real* root between 1 and 2, but no rational root.*

*If two integers p and q are relatively prime, their squares are relatively prime. Therefore $p^2 = 2q^2$ would imply that q^2 divided a number relatively prime to itself, which would imply $q^2 = \pm 1$, that is, $p^2 = \pm 2$, which is obviously impossible.

8 The Spectral Theorem

Theorem. *Let S be a symmetric matrix whose entries are real numbers. Then S has a spectral representation $S = a_1 P_1 + a_2 P_2 + \cdots + a_k P_k$, where a_1, a_2, ..., a_k are distinct real numbers and where $I = P_1 + P_2 + \cdots + P_k$ is an orthogonal partition of unity consisting of matrices whose entries are real numbers. Moreover, S determines both the numbers a_1, a_2, ..., a_k and the orthogonal projections P_i corresponding to the a_i.*

The numbers $a_1, a_2, \ldots, a_k$, called the **eigenvalues** of S, are the roots of the minimum polynomial of S, or, what is the same, the roots of the characteristic polynomial of S. (By the Cayley-Hamilton theorem, the characteristic polynomial is a multiple of the minimum polynomial, so any root of the minimum polynomial is a root of the characteristic polynomial. On the other hand, the characteristic polynomial is the determinant of a strongly diagonal matrix whose diagonal entries all divide the minimum polynomial, so the characteristic polynomial divides a power of the minimum polynomial, which shows that a root of the characteristic polynomial is a root of the minimum polynomial.)

Proof. The proofs of Section 5 and Section 6 apply—the only change being that the numbers involved are now real numbers—and show that the minimum polynomial of S changes sign a number of times equal to its degree. From the intermediate value property of real numbers, it then follows that the minimum polynomial of S has a number of real roots equal to its degree, so the minimum polynomial of S is of the form $f(x) = (x - a_1)(x - a_2) \cdots (x - a_k)$, where the a_i are distinct real numbers. The construction of Section 4 then gives the unique spectral representation of S.

9 Matrix Inversion

As was seen in Chapter 7, if a matrix A is invertible, then its mate $(A^T A)^{-1} A^T$ is its inverse. Therefore, A can be inverted if the *symmetric* matrix $A^T A$ can be inverted. (More generally, if A has a left inverse, then the mate $(A^T A)^{-1} A^T$ of A can be found by inverting the symmetric matrix $A^T A$, and if A has a right inverse then its mate $A^T (A A^T)^{-1}$ can be found by inverting the symmetric matrix $A A^T$.) One key to the importance of the spectral theorem is its use in the inversion of symmetric matrices.

Theorem. *Let $S = a_1 P_1 + a_2 P_2 + \cdots + a_k P_k$ be a spectral representation of a symmetric matrix S. If one of the coefficients a_i is zero, then S is not invertible. Otherwise, S is invertible and its inverse is $a_1^{-1} P_1 + a_2^{-1} P_2 + \cdots + a_k^{-1} P_k$.*

Proof. The minimum polynomial of S is the product $f(x)$ of the distinct linear polynomials $x - a_i$. If an a_i is zero, then the constant term of $f(x)$ is zero, say $f(x) = h(x)x$; if S were invertible then multiplication of $I = SS^{-1}$ by $h(S)$ would give $h(S) = h(S)SS^{-1} = f(S)S^{-1} = 0 \cdot S^{-1} = 0$, which is impossible because the degree of $h(x)$ is less than the degree of the minimum polynomial $f(x)$. Otherwise, because $P_i P_j = 0$ when $i \neq j$, the product of $a_1 P_1 + a_2 P_2 + \cdots + a_k P_k$ and $a_1^{-1} P_1 + a_2^{-1} P_2 + \cdots + a_k^{-1} P_k$ is $a_1 a_1^{-1} P_1^2 + a_2 a_2^{-1} P_2^2 + \cdots + a_k a_k^{-1} P_k^2 = P_1 + P_2 + \cdots + P_k = I$, as was to be shown.

10 *Diagonalizing Symmetric Matrices*

Because the minimum polynomial of a symmetric matrix is a product of distinct factors $x - a_i$ in which the a_i are real numbers, a symmetric matrix with rational or real entries is diagonalizable using real numbers. In other words, given a symmetric matrix S whose entries are real numbers, there is an invertible matrix U whose entries are real numbers such that $U^{-1}SU$ is diagonal. The usefulness of this remark is greatly enhanced by showing that the matrix U can be chosen in such a way that it is particularly easy to invert. Specifically, there is a U with this property whose inverse is simply its transpose.

Theorem. *Given a symmetric matrix S of real numbers, there is a matrix U of real numbers for which $U^T SU$ is diagonal and $U^T U = I$.*

Proof. Such a matrix U can be constructed in the following way. Let $S = a_1 P_1 + a_2 P_2 + \cdots + a_k P_k$ be a spectral representation of S. By the theorem of Section 2, the orthogonal partition of unity $I_n = P_1 + P_2 + \cdots + P_k$ can be "refined" to an orthogonal partition of unity $I_n = Q_1 + Q_2 + \cdots + Q_n$ in which each Q_i has rank 1 and each P is a sum of one or more Qs. Then $S = b_1 Q_1 + b_2 Q_2 + \cdots + b_n Q_n$ where each b is equal to one of the as.

For each $i = 1, 2, \ldots, n$, let C_i be a nonzero column of Q_i and let C be the $n \times n$ matrix whose ith column is C_i. Because $Q_i^2 = Q_i$, each C_i satisfies $Q_i C_i = C_i$. The entry in row i of column j of $C^T C$ is $C_i^T C_j = (Q_i C_i)^T (Q_j C_j) = C_i^T Q_i Q_j C_j$, which is zero when $i \neq j$ because then $Q_i Q_j = 0$. In other words, $C^T C$ is a diagonal matrix, call it D. The ith diagonal entry d_i of D is $C_i^T C_i$, which is *positive.* Let V be the diagonal matrix whose ith diagonal entry is $1/\sqrt{d_i}$. It will be shown that $U = CV$ has the two required properties.

First, with this U, $U^TU = V^TC^TCV = VDV = DV^2$ because V, being diagonal, is symmetric and commutes with the diagonal matrix D; thus $U^TU = DV^2 = I_n$. Since $SQ_i = b_iQ_i$, multiplying C on the left by S multiplies the ith column of C by b_i, which is the same as multiplying C on the right by the diagonal matrix B whose ith diagonal entry is b_i. In short, $SC = CB$. Therefore, $U^TSU = V^TC^TSCV = V^TC^TCBV = VDBV$ is a product of diagonal matrices, which shows not only that U^TSU is diagonal but also shows that it is equal to $DV^2B = B$.

Note that the number $\sqrt{d_i}$ is normally irrational. Therefore, the construction of U is not normally possible with rational numbers, even in cases in which S is diagonalizable over the rational numbers.

This theorem is often stated in the form: "There is an orthonormal basis of eigenvectors of S." To say the columns of U are a "basis" means that U is invertible. To say the columns of U are "orthonormal" means that $U^TU = I$. To say the columns of U are "eigenvectors" of S means that $SU = UB$ for some diagonal matrix B.

Examples

As is noted in Section 2,

$$\begin{bmatrix} 1 & 0 & 0 \\ 0 & 1 & 0 \\ 0 & 0 & 1 \end{bmatrix} = \begin{bmatrix} 1 & 0 & 0 \\ 0 & 0 & 0 \\ 0 & 0 & 0 \end{bmatrix} + \begin{bmatrix} 0 & 0 & 0 \\ 0 & 1 & 0 \\ 0 & 0 & 0 \end{bmatrix} + \begin{bmatrix} 0 & 0 & 0 \\ 0 & 0 & 0 \\ 0 & 0 & 1 \end{bmatrix}$$

is a 3×3 orthogonal partition of unity. A less obvious example can be obtained using the method of Section 2. For example, with $A = \begin{bmatrix} 1 \\ 2 \\ 1 \end{bmatrix}$, one finds the orthogonal projection

$$\begin{bmatrix} 1 \\ 2 \\ 1 \end{bmatrix} [6]^{-1} [1 \quad 2 \quad 1] = \begin{bmatrix} \frac{1}{6} & \frac{2}{6} & \frac{1}{6} \\ \frac{2}{6} & \frac{4}{6} & \frac{2}{6} \\ \frac{1}{6} & \frac{2}{6} & \frac{1}{6} \end{bmatrix},$$

call it P. With $Q = I - P$ one then finds the orthogonal partition of unity

$$I = P + Q = \begin{bmatrix} \frac{1}{6} & \frac{2}{6} & \frac{1}{6} \\ \frac{2}{6} & \frac{4}{6} & \frac{2}{6} \\ \frac{1}{6} & \frac{2}{6} & \frac{1}{6} \end{bmatrix} + \begin{bmatrix} \frac{5}{6} & -\frac{2}{6} & -\frac{1}{6} \\ -\frac{2}{6} & \frac{2}{6} & -\frac{2}{6} \\ -\frac{1}{6} & -\frac{2}{6} & \frac{5}{6} \end{bmatrix}.$$

The trace of Q is $\frac{5}{6} + \frac{2}{6} + \frac{5}{6} = 2$, so Q has rank 2 and the method of Section 2 can be used to refine the partition $I = P + Q$ further. Let A be a column of Q, say the middle column, (more generally, one can choose A to be any nonzero 3×1 matrix of the form QX, where X is a 3×1 matrix), and set

$$Q_1 = AA^M = \begin{bmatrix} -\frac{1}{3} \\ \frac{1}{3} \\ -\frac{1}{3} \end{bmatrix} [\tfrac{3}{9}]^{-1} [-\tfrac{1}{3} \quad \tfrac{1}{3} \quad -\tfrac{1}{3}] = \begin{bmatrix} \frac{1}{3} & -\frac{1}{3} & \frac{1}{3} \\ -\frac{1}{3} & \frac{1}{3} & -\frac{1}{3} \\ \frac{1}{3} & -\frac{1}{3} & \frac{1}{3} \end{bmatrix}.$$

Then, with $Q_2 = Q - Q_1$,

$$\begin{aligned} I_3 &= P + Q_1 + Q_2 \\ &= \begin{bmatrix} \frac{1}{6} & \frac{1}{3} & \frac{1}{6} \\ \frac{1}{3} & \frac{2}{3} & \frac{1}{3} \\ \frac{1}{6} & \frac{1}{3} & \frac{1}{6} \end{bmatrix} + \begin{bmatrix} \frac{1}{3} & -\frac{1}{3} & \frac{1}{3} \\ -\frac{1}{3} & \frac{1}{3} & -\frac{1}{3} \\ \frac{1}{3} & -\frac{1}{3} & \frac{1}{3} \end{bmatrix} + \begin{bmatrix} \frac{1}{2} & 0 & -\frac{1}{2} \\ 0 & 0 & 0 \\ -\frac{1}{2} & 0 & \frac{1}{2} \end{bmatrix} \end{aligned} \tag{1}$$

is an orthogonal partition of unity, as is easily checked. Note that each projection in this partition of unity has trace 1.

The two choices in this construction—of the nonzero 3×1 matrix $\begin{bmatrix} 1 \\ 2 \\ 1 \end{bmatrix}$ at the first step and of the nonzero 3×1 matrix $A = Q\begin{bmatrix} 0 \\ 1 \\ 0 \end{bmatrix}$ at the second step—can be altered to find many other orthogonal partitions of I_3, or, by the same method, of I_n for any $n > 1$.

If the three projections in the partition of unity (1) are multiplied respectively by 6, 3, and 2 and added, the result is

$$\begin{aligned} S &= 6\begin{bmatrix} \frac{1}{6} & \frac{1}{3} & \frac{1}{6} \\ \frac{1}{3} & \frac{2}{3} & \frac{1}{3} \\ \frac{1}{6} & \frac{1}{3} & \frac{1}{6} \end{bmatrix} + 3\begin{bmatrix} \frac{1}{3} & -\frac{1}{3} & \frac{1}{3} \\ -\frac{1}{3} & \frac{1}{3} & -\frac{1}{3} \\ \frac{1}{3} & -\frac{1}{3} & \frac{1}{3} \end{bmatrix} + 2\begin{bmatrix} \frac{1}{2} & 0 & -\frac{1}{2} \\ 0 & 0 & 0 \\ -\frac{1}{2} & 0 & \frac{1}{2} \end{bmatrix} \\ &= \begin{bmatrix} 3 & 1 & 1 \\ 1 & 5 & 1 \\ 1 & 1 & 3 \end{bmatrix}. \end{aligned} \tag{2}$$

The characteristic polynomial of this S can be found without too much difficulty to be $x^3 - 11x^2 + 36x - 36$ (see Exercise 5 of Chapter 9 or the Examples section of Chapter 8). Its roots are easily found (if the roots of a polynomial are all rational, they are always easy to find) to be 6, 3, and 2, and the polynomial itself is found to be $(x - 6)(x - 3)(x - 2)$. Since it has no repeated factors, this polynomial is also the minimum polynomial of S and the theorem of Section 4 implies that $S = 6P_1 + 3P_2 + 2P_3$ for some orthogonal partition of unity $I = P_1 + P_2 + P_3$. By the construction in the proof of the theorem of Section 4, P_1 can be found by first computing $G_1 = (S - 3I)(S - 2I) = \begin{bmatrix} 0 & 1 & 1 \\ 1 & 2 & 1 \\ 1 & 1 & 0 \end{bmatrix}\begin{bmatrix} 1 & 1 & 1 \\ 1 & 3 & 1 \\ 1 & 1 & 1 \end{bmatrix} = \begin{bmatrix} 2 & 4 & 2 \\ 4 & 8 & 4 \\ 2 & 4 & 2 \end{bmatrix}$.

Thus $G_1 = \begin{bmatrix} 2 \\ 4 \\ 2 \end{bmatrix} [1 \quad 2 \quad 1]$, so $H_1 = [1 \quad 2 \quad 1]^M \begin{bmatrix} 2 \\ 4 \\ 2 \end{bmatrix}^M$ and

$$\begin{aligned} P_1 &= \begin{bmatrix} 2 \\ 4 \\ 2 \end{bmatrix} [1 \quad 2 \quad 1][1 \quad 2 \quad 1]^M \begin{bmatrix} 2 \\ 4 \\ 2 \end{bmatrix}^M \\ &= \begin{bmatrix} 2 \\ 4 \\ 2 \end{bmatrix} \begin{bmatrix} 2 \\ 4 \\ 2 \end{bmatrix}^M = \begin{bmatrix} 2 \\ 4 \\ 2 \end{bmatrix} \left[\frac{1}{2^2+4^2+2^2} \right] [2 \quad 4 \quad 2], \end{aligned}$$

which is easily seen to be the matrix that is multiplied by 6 in (2) above. The matrices P_2 and P_3 can of course be found in the same way (but, like P_1, they are already known from equation (2)).

Note that the *inverse* of the matrix in (2) can be found using the formula

$$\begin{aligned} S^{-1} &= \frac{1}{6} \begin{bmatrix} \frac{1}{6} & \frac{1}{3} & \frac{1}{6} \\ \frac{1}{3} & \frac{2}{3} & \frac{1}{3} \\ \frac{1}{6} & \frac{1}{3} & \frac{1}{6} \end{bmatrix} + \frac{1}{3} \begin{bmatrix} \frac{1}{3} & -\frac{1}{3} & \frac{1}{3} \\ -\frac{1}{3} & \frac{1}{3} & -\frac{1}{3} \\ \frac{1}{3} & -\frac{1}{3} & \frac{1}{3} \end{bmatrix} + \frac{1}{2} \begin{bmatrix} \frac{1}{2} & 0 & -\frac{1}{2} \\ 0 & 0 & 0 \\ -\frac{1}{2} & 0 & \frac{1}{2} \end{bmatrix} \\ &= \begin{bmatrix} \frac{14}{36} & -\frac{2}{36} & -\frac{4}{36} \\ -\frac{2}{36} & \frac{8}{36} & -\frac{2}{36} \\ -\frac{4}{36} & -\frac{2}{36} & \frac{14}{36} \end{bmatrix}. \end{aligned}$$

That this matrix is indeed the inverse of S is easy to check.

The characteristic polynomial of the symmetric matrix $S = \begin{bmatrix} 1 & 1 \\ 1 & 2 \end{bmatrix}$ is $(x-1)(x-2) - (-1)^2 = x^2 - 3x + 1$. Since this polynomial is irreducible, it is the minimum polynomial $f(x)$ of S. That this polynomial has a number of sign changes equal to its degree follows simply from the observation that $f(0) = 1$, $f(2) = -1$, $f(4) = 5$. The general proof in Section 6 that this polynomial has two sign changes applies to this example in the following way. Because $\operatorname{tr}(S^0) = 2$, $\operatorname{tr}(S^1) = 3$, and $\operatorname{tr}(S^2) = 7$, the matrix B is $\begin{bmatrix} 7 & 3 \\ 3 & 2 \end{bmatrix}$. Therefore, $H = \begin{bmatrix} \frac{2}{5} & -\frac{3}{5} \\ -\frac{3}{5} & \frac{7}{5} \end{bmatrix} \begin{bmatrix} 1 \\ 0 \end{bmatrix} = \begin{bmatrix} \frac{2}{5} \\ -\frac{3}{5} \end{bmatrix}$, so $h(x) = \frac{2}{5}x - \frac{3}{5}$. One can then find $1 = (-\frac{6}{5}x + 1)f(x) + (3x^2 - 7x)h(x)$, which shows that $s_1(x)$ is the remainder when $3x^2 - 7x$ is divided by $f(x)$. Thus $s_2(x) = x^2 - 3x + 1$ and $s_1(x) = 2x - 3$, so $s_0(x)$, which is the negative of the remainder when $s_2(x)$ is divided by $s_1(x)$, is $\frac{5}{4}$. This sequence of three polynomials satisfies conditions (i) and (ii) of the Theorem of Section 5, so $s_2(x) = x^2 - 3x + 1$ must have two sign changes. (Note that in this case $s_1(x)$ is the derivative of $s_2(x)$—see Exercise 12. Note also that $J(0) = 2$, $J(2) = 1$, and $J(4) = 0$.)

Therefore, $x^2 - 3x + 1$ has two irrational roots (the polynomial is monic, so a rational root would have to be an integer), call them a_1 and a_2, and, by the spectral theorem, $S = a_1 P_1 + a_2 P_2$ for some partition of unity $I = P_1 + P_2$.

Now, in the notation of the proof of Section 4, $G_1 = S - a_2 I = \begin{bmatrix} 1-a_2 & 1 \\ 1 & 2-a_2 \end{bmatrix}$. Because G_1 is not invertible, its determinant must be zero, and indeed the second column is $2 - a_2$ times the first, because $(2-a_2)(1-a_2) = a_2^2 - 3a_2 + 2 = 1$. Thus $G_1 = \begin{bmatrix} 1-a_2 \\ 1 \end{bmatrix} [\, 1 \quad 2-a_2 \,]$ from which $H_1 = [\, 1 \quad 2-a_2 \,]^M \begin{bmatrix} 1-a_2 \\ 1 \end{bmatrix}^M$ and

$$\begin{aligned} P_1 &= \begin{bmatrix} 1-a_2 \\ 1 \end{bmatrix} [\, 1 \quad 2-a_2 \,][\, 1 \quad 2-a_2 \,]^M \begin{bmatrix} 1-a_2 \\ 1 \end{bmatrix}^M = \begin{bmatrix} 1-a_2 \\ 1 \end{bmatrix} \begin{bmatrix} 1-a_2 \\ 1 \end{bmatrix}^M \\ &= \begin{bmatrix} 1-a_2 \\ 1 \end{bmatrix} \frac{1}{(1-a_2)^2+1^2} [\, 1-a_2 \quad 1 \,] = \frac{1}{a_2^2-2a_2+2} \begin{bmatrix} (1-a_2)^2 & 1-a_2 \\ 1-a_2 & 1 \end{bmatrix}. \end{aligned}$$

The denominator of the fraction in front of this matrix can be rationalized by observing that $a_2^2 - 2a_2 + 2 = a_2^2 - 3a_2 + 1 + a_2 + 1 = a_2 + 1$ and $(-a_2+4)(a_2+1) = -a_2^2 + 3a_2 + 4 = 5$, so the factor in front is $\frac{-a_2+4}{5}$. Thus

$$P_1 = \frac{-a_2+4}{5} \begin{bmatrix} a_2 & -a_2+1 \\ -a_2+1 & 1 \end{bmatrix} = \begin{bmatrix} \frac{a_2+1}{5} & \frac{-2a_2+3}{5} \\ \frac{-2a_2+3}{5} & \frac{-a_2+4}{5} \end{bmatrix}$$

as easy computations show. Replacement of a_2 by a_1 throughout this calculation gives

$$P_2 = \begin{bmatrix} \frac{a_1+1}{5} & \frac{-2a_1+3}{5} \\ \frac{-2a_1+3}{5} & \frac{-a_1+4}{5} \end{bmatrix}.$$

The equations $S = a_1 P_1 + a_2 P_2$, $I = P_1 + P_2$, $P_1^2 = P_1$, $P_2^2 = P_2$, and $P_1 P_2 = 0$ are easily checked using the identities $a_1 + a_2 = 3$ and $a_1 a_2 = 1$ which follow from $(x-a_1)(x-a_2) = x^2 - 3x + 1$.

The solution can be expressed in terms of explicit real numbers by using the quadratic formula to write the roots as $a_1 = \frac{3+\sqrt{5}}{2}$ and $a_2 = \frac{3-\sqrt{5}}{2}$, which gives

$$S = \frac{3+\sqrt{5}}{2} \begin{bmatrix} \frac{5-\sqrt{5}}{10} & \frac{\sqrt{5}}{5} \\ \frac{\sqrt{5}}{5} & \frac{5+\sqrt{5}}{5} \end{bmatrix} + \frac{3-\sqrt{5}}{2} \begin{bmatrix} \frac{5+\sqrt{5}}{10} & -\frac{\sqrt{5}}{5} \\ -\frac{\sqrt{5}}{5} & \frac{5-\sqrt{5}}{5} \end{bmatrix}$$

and

$$I = \begin{bmatrix} \frac{5-\sqrt{5}}{10} & \frac{\sqrt{5}}{5} \\ \frac{\sqrt{5}}{5} & \frac{5+\sqrt{5}}{5} \end{bmatrix} + \begin{bmatrix} \frac{5+\sqrt{5}}{10} & -\frac{\sqrt{5}}{5} \\ -\frac{\sqrt{5}}{5} & \frac{5-\sqrt{5}}{5} \end{bmatrix}.$$

Application of the construction of Section 8 to this example gives, for example,

$$C = \begin{bmatrix} \frac{5-\sqrt{5}}{10} & \frac{5+\sqrt{5}}{10} \\ \frac{\sqrt{5}}{5} & -\frac{\sqrt{5}}{5} \end{bmatrix}$$

when the first column of $P_1 = Q_1$ is chosen as the first column of C and the first column of $P_2 = Q_2$ is chosen as the second column of C. Then $C^T C = D$ where

D is the 2×2 diagonal matrix with diagonal entries $d_1 = \left(\frac{5-\sqrt{5}}{10}\right)^2 + \left(\frac{\sqrt{5}}{5}\right)^2 = \frac{5-\sqrt{5}}{10}$ and $d_2 = \frac{5+\sqrt{5}}{10}$. Thus,

$$U^T U = I_2 \qquad \text{and} \qquad U^T \begin{bmatrix} 1 & 1 \\ 1 & 2 \end{bmatrix} U = \begin{bmatrix} \frac{3+\sqrt{5}}{2} & 0 \\ 0 & \frac{3-\sqrt{5}}{2} \end{bmatrix}$$

when U is the matrix

$$U = CV = \begin{bmatrix} \frac{5-\sqrt{5}}{10} & \frac{5+\sqrt{5}}{10} \\ \frac{\sqrt{5}}{5} & -\frac{\sqrt{5}}{5} \end{bmatrix} \begin{bmatrix} \left(\frac{5-\sqrt{5}}{10}\right)^{-\frac{1}{2}} & 0 \\ 0 & \left(\frac{5+\sqrt{5}}{10}\right)^{-\frac{1}{2}} \end{bmatrix} = \begin{bmatrix} a & b \\ c & d \end{bmatrix}$$

where a and b are the positive square roots of $\frac{5-\sqrt{5}}{10}$ and $\frac{5+\sqrt{5}}{10}$, respectively, and $c = \frac{\sqrt{5}}{5a} = \frac{a\sqrt{5}}{5 \cdot \frac{5-\sqrt{5}}{10}} = \frac{2a\sqrt{5}}{5-\sqrt{5}} = \frac{2a\sqrt{5}(5+\sqrt{5})}{25-5} = \frac{2a(5\sqrt{5}+5)}{20} = \frac{a(\sqrt{5}+1)}{2}$ and $d = -\frac{b(\sqrt{5}-1)}{2}$. The four irrational numbers a, b, c, and d defined in this way have the required properties, as can be checked using a calculator.

Exercises

1 The following symmetric matrices S are diagonalizable. Find their spectral representations.

(a) $\begin{bmatrix} 2 & 2 \\ 2 & -1 \end{bmatrix}$ (b) $\begin{bmatrix} 17 & -6 \\ -6 & 22 \end{bmatrix}$ (c) $\begin{bmatrix} 6 & -2 & -3 \\ -2 & 3 & -6 \\ -3 & -6 & -2 \end{bmatrix}$ (d) $\begin{bmatrix} 25 & 10 & -38 \\ 10 & -38 & 25 \\ -38 & 25 & 10 \end{bmatrix}$

2 Let a and b be rational numbers, not both zero. Find the unique 2×2 partition of unity $I_2 = P+Q$ into projections of rank 1 for which $P\begin{bmatrix} a \\ b \end{bmatrix} = \begin{bmatrix} a \\ b \end{bmatrix}$.

3 Check that

$$\begin{bmatrix} 1 & 0 & 0 & 0 \\ 0 & 1 & 0 & 0 \\ 0 & 0 & 1 & 0 \\ 0 & 0 & 0 & 1 \end{bmatrix} = \begin{bmatrix} \frac{3}{5} & \frac{1}{5} & \frac{1}{5} & -\frac{2}{5} \\ \frac{1}{5} & \frac{2}{5} & \frac{2}{5} & \frac{1}{5} \\ \frac{1}{5} & \frac{2}{5} & \frac{2}{5} & \frac{1}{5} \\ -\frac{2}{5} & \frac{1}{5} & \frac{1}{5} & \frac{3}{5} \end{bmatrix} + \begin{bmatrix} \frac{2}{5} & -\frac{1}{5} & -\frac{1}{5} & \frac{2}{5} \\ -\frac{1}{5} & \frac{3}{5} & -\frac{2}{5} & -\frac{1}{5} \\ -\frac{1}{5} & -\frac{2}{5} & \frac{3}{5} & -\frac{1}{5} \\ \frac{2}{5} & -\frac{1}{5} & -\frac{1}{5} & \frac{2}{5} \end{bmatrix}$$

is an orthogonal partition of I_4. Find an orthogonal partition of I_4 containing four terms which is a refinement of this one in the sense that each of the two terms of this one is a sum of terms of the refinement (see Section 2).

4 Begin with the four-term partition of I_4 of the preceding exercise, choose distinct rational numbers a_1, a_2, a_3, a_4, and compute $a_1P_1 + a_2P_2 + a_3Q_1 + a_4Q_2$, call it S. Then use the method of Section 4 to find the spectral representation of S, that is, to reconstruct the as and $I = P_1 + P_2 + Q_1 + Q_2$

5 Construct an orthogonal partition $I_3 = P_1 + P_2 + P_3$, choose three distinct rational numbers a_1, a_2, a_3, and use the method of Section 4 to reconstruct the as and Ps given just $S = a_1P_1 + a_2P_2 + a_3P_3$.

6 Making use of the last example in the Examples section, give a geometrical description of the curves $x^2 + 2xy + 2y^2 =$ constant in the xy-plane. [Regard the matrix U as a *change of coordinates* and interpret the equations $U^TU = I$ and $U^TST =$ diagonal geometrically.]

7 Find the spectral representation of $S = \begin{bmatrix} 30 & -8 & -10 \\ -8 & 37 & 4 \\ -10 & 4 & 15 \end{bmatrix}$ and use this representation, as in the preceding exercise, to give a geometrical description of the surfaces $30x^2 + 37y^2 + 15z^2 - 16xy - 20xz + 8yz =$ constant.

8 In connection with the pattern of the signs of the polynomials $s_i(x)$ in Section 5, show that there is only one way that the pattern of signs $+ - +$ can change to the pattern of signs $+ + +$ when changes obey the rules: (1) A sign other than the last changes only when its neighboring signs are opposite; (2) when a sign changes, it changes from disagreement to agreement with the preceding sign; and (3) the first sign is always $+$. Show that there are *two* ways for the pattern $+ - + -$ to change to $+ + + +$ while respecting these rules.

9 Prove that if $f(x)$ and $g(x)$ are given polynomials, and if $g(x) \neq 0$, there exist unique polynomials $q(x)$ and $r(x)$ such that $f(x) = q(x)g(x) + r(x)$ and $\deg r < \deg g$.

10 Prove that a polynomial of degree n changes sign at most n times.

11 Find the minimum polynomial of the symmetric matrix $S = \begin{bmatrix} -2 & -1 & 0 \\ -1 & 1 & 2 \\ 0 & 2 & 3 \end{bmatrix}$. Show that it has a number of sign changes equal to its degree. Find a sequence of polynomials that starts with this polynomial and satisfies the conditions of the theorem of Section 5.

12 Let S be a symmetric matrix, let $f(x)$ be its minimum polynomial, and let $S = a_1 P_1 + a_2 P_2 + \cdots + a_k P_k$ be its spectral representation. (The numbers a_i need not be rational.) Prove that when the matrices P_i all have rank 1 (or, what is the same, all have trace 1) the polynomial $s_{m-1}(x)$ constructed in Section 6 is simply the derivative of $f(x)$. (The derivative of a polynomial $f(x)$ is the coefficient of h in the polynomial in two variables $f(x+h)$. Note that this is a strictly algebraic notion and has nothing to do with limits.) Prove that in any case $s_{m-1}(x)$ is the unique polynomial of degree less than $m = \deg f(x)$ for which $s_{m-1}(a_i) = r_i f'(a_i)$, where $f'(x)$ is the derivative of $f(x)$ and r_i is the trace of P_i.

Appendix

Linear Programming

1 The Problem

A LINEAR PROGRAMMING PROBLEM is a problem of maximizing a linear polynomial subject to a set of linear constraints which may be either equalities or inequalities. A linear polynomial here is, more precisely, an inhomogeneous linear polynomial $c_1x_1 + c_2x_2 + \cdots + c_nx_n + d$, where $c_1, c_2, \ldots, c_n, d$ are rational numbers and $x_1, x_2, \ldots, x_n$ are the variables of the problem. A linear constraint is a condition of the form $g(x) \geq h(x)$ or $g(x) = h(x)$ on the values of the variables, where g and h are linear polynomials in $x_1, x_2, \ldots, x_n$.

The word "linear" refers, of course, to the linear polynomials in the statement of the problem. The word "programming" has a less obvious origin. The subject came into being in the late 1940s, when the newly developed electronic computers were first being applied to the analysis of large and complex systems like factories or fleets of ships. Thus the word refers not to the programming of a computer or to the method by which the problem is solved, but rather to the solution itself, which is the optimum "program" for the operation of the system that the problem is designed to represent.

Since the terms in $g(x) \geq h(x)$ or $g(x) = h(x)$ containing variables can all be moved to one side and the constant terms can all be moved to the other, such a constraint can always be written $a_1x_1 + a_2x_2 + \cdots + a_nx_n \leq b$ or $a_1x_1 + a_2x_2 + \cdots + a_nx_n = b$ where $a_1, a_2, \ldots, a_n$, and b are rational numbers.

Note that the minimization of a linear polynomial f subject to such a set of constraints is the same as the maximization of $-f$ subject to the same constraints, so there is no loss of generality in restricting consideration to maximization problems. However, the key to the solution of linear programming problems lies in an investigation of the *constraints*, not the function to be maximized. Therefore, the next few sections deal with the constraints.

2 Standard Form for the Constraints

Any set of linear constraints of the type described in Section 1 can be restated as a set in which all inequality constraints have the special form $x_i \geq 0$, because any other inequality constraint $a_1x_1 + a_2x_2 + \cdots + a_nx_n \leq b$, can be stated as an equation and an inequality: $a_1x_1 + a_2x_2 + \cdots + a_nx_n + u = b$ and $u \geq 0$, where u is a new variable. (If this procedure is applied to a constraint already of the form $x_i \geq 0$, it first rewrites the constraint as $-x_i \leq 0$ and then sets $-x_i + u = 0$ where $u \geq 0$. Thus, $x_i \geq 0$ becomes $x_i = u$, where $u \geq 0$, which accomplishes nothing.) A variable x_i subject to the constraint $x_i \geq 0$ will be called a **special variable**. Thus, any set of equality and inequality constraints can be restated as a set in which the only constraints are equations and requirements that certain variables be special.

Once the constraints have been put in this form, any variable that is not special can be eliminated. For example, if x_1 is not special, one equation containing x_1 can be used to express x_1 as a linear polynomial in the other variables. This expression can then be used to eliminate x_1 from all other equality constraints and from the function to be maximized, without changing the inequality constraints (none of which involve x_1). Then x_1 plays no role in the statement of the problem and can be dropped. Therefore, *all* variables can be assumed, without loss of generality, to be special.

Constraints of this type—say they involve m equality constraints on n special variables—can be written in matrix form as "$AX = B$, $X \geq 0$," where X is an $n \times 1$ matrix whose entries are the variables $x_1, x_2, \ldots, x_n$ of the problem, where A is an $m \times n$ matrix of rational numbers, where B is an $m \times 1$ matrix of rational numbers, and where $X \geq 0$ is an abbreviated way of saying that all variables are special.

A **feasible point** of the constraints $AX = B$, $X \geq 0$ is a set of values satisfying them, that is, n nonnegative rational values for the n variables x_1, x_2, $\ldots$, x_n that satisfy the m equations $AX = B$. As the example $x_1 + x_2 = -1$, $x_1 \geq 0$, $x_2 \geq 0$ shows, there may be *no* feasible points for a given set of constraints. When this is the case, the constraints are said to be **contradictory**.

Let constraints be given in the form $AX = B$, $X \geq 0$, and let M and N be unimodular matrices such that MAN is strongly diagonal, say $MAN = D$. Since M is invertible, $AX = B$ if and only if $MAX = MB$. If the rank r of A is less than m, the last $m - r$ rows of D are zero, so the last $m - r$ rows of $DN^{-1} = MA$ are zero. The right-hand sides of the last $m - r$ equations $MAX = MB$ are therefore zero, so, unless the constraints are contradictory, the last $m - r$ entries of MB must be zero as well. When this is the case, the last $m - r$ equations of $MAX = MB$ are all the vacuous statement that $0 = 0$, so the equations $MAX = MB$ are equivalent to just the first r equations of $MAX = MB$. With A' equal to the first r rows of MA and B' equal to the first r rows of MB, it follows that, unless the constraints are contradictory, the equations $AX = B$ are equivalent to the equations $A'X = B'$. The rank of this A' is the same as the rank of MA (the omitted rows are all zero), which is r (MA is equivalent to A); that is, the rank of A' is equal to the number of its rows. Therefore, unless the constraints $AX = B$, $X \geq 0$ are contradictory, they can be replaced by another, equivalent set of constraints $A'X = B'$, $X \geq 0$ of the same form in which the rank of A' is equal to the number of its rows. In short, the constraints in a linear programming problem can be assumed without loss of generality (because nothing is lost if contradictory constraints are excluded) to be of the form $AX = B$, $X \geq 0$ where the rank of A is equal to the number of its rows.

For these reasons, the constraints of a linear programming problem will be said to be in **standard form** if they are of the form "$AX = B$, $X \geq 0$," where A is an $m \times n$ matrix of rank m and B is an $m \times 1$ matrix, both with rational entries.

The objective is to find, given such a set of constraints and given a linear polynomial f in the same variables $x_1, x_2, \ldots, x_n$, the largest value f can attain for rational values of the variables that satisfy the constraints. (The solution will take into account the two other possibilities: the constraints may be contradictory, or the constraints may not imply any upper bound on f.)

3 *Vertices*

The assumption that the $m \times n$ matrix A has rank m implies that the equations $AX = B$ can be "solved" for some set of m of the xs in terms of the remaining $n - m$. Specificially, A has a nonzero $m \times m$ minor, so $n - m$ columns of A can be deleted in such a way that the remaining $m \times m$ matrix is invertible. Say columns $i_1, i_2, \ldots, i_m$ of A contain an invertible matrix. To simplify the notation, let the xs be rearranged so that $i_1, i_2, \ldots, i_m$ become $1, 2, \ldots, m$. When T is the inverse of the invertible matrix in the first m columns of A, the constraints $TAX = TB$, $X \geq 0$, which are obviously equivalent to $AX = B$, $X \geq 0$, have the form $x_i + \alpha_{i,m+1}x_{m+1} + \cdots + \alpha_{i,n}x_n = \beta_i$ ($i = 1, 2, \ldots, m$), $x_i \geq 0$ ($i = 1, 2, \ldots, n$), where α_{ij} is the entry in row i and column j of TA and β_i is the entry in row i of TB. The solutions of $AX = B$ are parameterized by the variables $x_{m+1}, x_{m+2}, \ldots, x_n$ in the sense that, when the values of these variables in a solution are given, the remaining values are determined by the m equations $x_i = \beta_i - \alpha_{i,m+1}x_{m+1} - \cdots - \alpha_{i,n}x_n$.

In the same way, if the matrix contained in columns $i_1, i_2, \ldots, i_m$ of A is invertible, and if its inverse is T, then the equations $TAX = TB$ express x_i for each i in the list $i_1, i_2, \ldots, i_m$ as a linear polynomial in the remaining $n - m$ variables and therefore give a parameterization of the solutions of $AX = B$ by $n-m$ variables. It would be reasonable to call the m variables *dependent* variables and the remaining $n - m$ *independent* variables, but the traditional terminology of linear programming calls the m variables *basic* and calls the $n - m$ variables *nonbasic*. We will adopt a hybrid terminology, calling the m variables **basic** variables and the remaining $n - m$ variables the **independent** variables.

We will define a **vertex** of a set of constraints $AX = B$, $X \geq 0$ in standard form to be an ordered list of m distinct indices $i_1, i_2, \ldots, i_m$ in the range $1 \leq i_j \leq n$ with the property that the $m \times m$ matrix whose columns are the listed columns of A is invertible. Given a vertex of $AX = B$, $X \geq 0$, its basic variables are the variables whose indices are listed, and its independent variables are the others.

This use of the word "vertex" for the choice of the basic variables is entirely artificial and should not be taken literally. A vertex, as just defined, does give rise to a "point" or set of values of the n variables of the problem, namely, the solution of $AX = B$ in which all the independent variables have the value 0. However, the vertex is a listing of the *other* variables, the basic variables, which may or may not be zero at the point, and it is an *ordering* of these variables, which certainly has nothing to do with the geometrical point.

We will say that a vertex is **made explicit** by the matrix T with the property that column i_j of TA is column j of I_m for $j = 1, 2, \ldots, m$. Note that T takes into account the order of the basic variables; solving $TAX = TB$ to express the basic variables in terms of the independent variables uses the first equation to express the first basic variable, the second equation to express the second, and so forth.

When constraints are in standard form $AX = B$, $X \geq 0$, any invertible $m \times m$ matrix M leads to an equivalent statement of the constraints, namely, $MAX = MB$, $X \geq 0$. The simplex method is, briefly put, an algorithm for generating a sequence of restatements of the constraints of a given linear programming problem in the form $T_iAX = T_iB$, $X \geq 0$, where the matrices T_1, T_2, T_3, ... make explicit a sequence of vertices of the contraints. The sequence terminates and the problem is solved when a form of the constraints is reached in which the solution of the problem is obvious.

4 Feasible Vertices

A vertex of the constraints $AX = B$, $X \geq 0$ in standard form parameterizes the solutions of $AX = B$ by describing how to find *all* values of the variables in a solution when just the values of the independent variables are given. A given set of values for the independent variables determines a feasible point of the constraints $AX = B$, $X \geq 0$ if and only if the values of all n variables determined in this way are nonnegative, which is to say that not only are the values of the $n - m$ independent variables nonnegative but so also are the values of the m linear polynomials that express the values of the basic variables in terms of the independent variables.

A **feasible vertex** of a set of constraints is a vertex with the property that the point obtained by setting the independent variables of the vertex equal to zero is a feasible point. A vertex of $AX = B$, $X \geq 0$ is feasible if and only if the matrix T that makes the vertex explicit has the property that all m entries of TB are nonnegative, because the entries of TB are the constant terms of the linear polynomials that express the basic variables in terms of the independent variables.

(The set of all feasible points can be imagined as being a polyhedron in an $n - m$ dimensional space parameterized by the independent variables. The faces of the polyhedron lie in the n hyperplanes $x_i = 0$, and its vertices lie at intersections of $n - m$ of these hyperplanes. A feasible vertex gives rise to a vertex of the polyhedron, namely, the point at which all *independent* variables are zero.)

5 *Solution Vertices*

Let a linear programming problem with a feasible vertex be given. What is given, then, is n variables $x_1, x_2, \ldots, x_n$; a linear polynomial in these variables to be maximized, call it $f = c_1x_1 + c_2x_2 + \cdots + c_nx_n + d$; an $m \times n$ matrix A of rank m with rational entries; an $m \times 1$ matrix B with rational entries; and an ordered set of m indices $i_1, i_2, \ldots, i_m$ specifying the choice of basic variables, such that giving the value zero to all the independent variables gives nonnegative values to all the basic variables.

Each of the m basic variables can be expressed as a linear polynomial in the independent variables. Substitution of these expressions in $f = c_1x_1 + c_2x_2 + \cdots + c_nx_n + d$ expresses f as a linear polynomial in the xs in which only the independent variables appear, say $f = C_1x_1 + C_2x_2 + \cdots + C_nx_n + D$, where $C_i = 0$ whenever x_i is a basic variable.

Setting all independent variables equal to zero gives a feasible point (by assumption) at which the value of f is D. *If no coefficient C_j is positive, the value D of f at the feasible point where the independent variables are all zero is its maximum value subject to the given constraints*, because the value of f at any point is equal to D plus the sum of the numbers C_jx_j, where j ranges over the indices of the independent variables and x_j is the value of that independent variable at the point in question; at a *feasible* point all variables are nonnegative, so $\sum C_j x_j \leq 0$ when no C_j is positive, that is, the value of f at a feasible point is at most D.

Therefore, we will call a feasible vertex at which no coefficient C_j is positive a **solution vertex** of the problem.

(We will ignore the problem of finding *all* feasible points at which the maximum value of f is attained. Briefly, if no C_j is zero, the point just found is the only one. Otherwise, the set of all such points can be described in terms of constraints on the independent variables of a solution vertex.)

6 The Simplex Method

The simplex method is an algorithm for finding, given a feasible vertex that is not a solution vertex, a better feasible vertex. The "better" feasible vertex is found as follows:

Because the given feasible vertex is not a solution vertex, at least one of the coefficients C_j described in the last section is positive. Let j be an index for which C_j is as large as possible. (In other words, let x_j be an independent variable for which the partial derivative $\partial f/\partial x_j$ is as large as possible.) In the equations $TAX = TB$, where T makes explicit the given feasible vertex, let all independent variables other than x_j be set equal to zero. The result is a set of m equations of the form $x_i + \alpha_i x_j = \beta_i$, where i ranges over the m indices of the basic variables. The constraint $x_i \geq 0$ implies $\alpha_i x_j \leq \beta_i$. When $\alpha_i \leq 0$, this constraint is a consequence of $x_j \geq 0$ ($\beta_i \geq 0$ because the given vertex is feasible). If $\alpha_i > 0$, on the other hand, $\alpha_i x_j \leq \beta_i$ is an upper-bound constraint $x_j \leq \beta_i/\alpha_i$ on x_j. If none of the coefficients α_i are positive, the linear programming problem is solved in the trivial sense that a description of the problem has been found in which f is clearly *not bounded above* by the constraints, because assigning x_j a large value K while assigning the other independent variables the value zero gives a feasible point at which f has the value $C_j K + D$, which can be made as large as one pleases. Otherwise, let i be the index with $\alpha_i > 0$ for which β_i/α_i is smallest. (The issue of what to do in case of a tie is discussed in Section 8. In practical applications of the simplex method, ties are broken by choosing arbitrarily one of the indices for which $\alpha_i > 0$ and β_i/α_i is smallest.) *The next feasible vertex is the one in which x_j replaces x_i as a basic variable.*

Setting all independent variables other than x_j equal to 0 while x_j varies in the range $0 \leq x_j \leq \beta_i/\alpha_i$ gives, by the choice of i, a line segment of feasible points. (It can happen that $\beta_i = 0$, in which case the line segment degenerates to a single point. In this case, the algorithm proceeds *as if* it were moving to a new point, even though it is only proceeding to a new description of the same point. The rationale for the algorithm in this case is considered in Section 8.) Because $C_j > 0$, the value of f *increases* as the value of x_j increases from 0 to β_i/α_i. Therefore, the value of f increases when one passes from the point where all independent variables of the original feasible vertex are zero (and x_i is β_i) to the point where all independent variables of the new feasible vertex are zero (and x_j is β_i/α_i).

That replacement of x_i with x_j gives a vertex is easy to see. The columns of TA that correspond to the new basic variables contain $m-1$ of the m columns of I_m and contain column j of TA. The missing column of I_m is the one in column i of TA, so it has its 1 in the row corresponding to x_i, and the entry α_i of column j of TA in this row is positive. Thus, α_i is the only nonzero entry in its row of the m columns of TA in question, and when this row is deleted from the other $m-1$ columns, the resulting columns are the columns of I_{m-1}. It follows that the determinant of these m columns of TA is $\pm\alpha_i \neq 0$, which implies that the determinant of these m columns of A is nonzero.

The simplex method applies this algorithm repeatedly to move from one feasible vertex to the next until a solution vertex is found (or until a statement of the problem is found that shows f assumes arbitrarily large values at feasible points). Note that the method assumes that a feasible vertex of the given constraints is known. The problem of finding a feasible vertex with which to begin is addressed in Section 9 below.

7 *Implementation of the Simplex Algorithm*

Let T be the matrix that makes the given feasible vertex explicit. The simplex algorithm is, in essence, an algorithm for finding the matrix T' that makes explicit the next feasible vertex. The construction of T' can be described explicitly as follows.

Once the index j for which C_j assumes its maximum value is known, compute column j of TA. This is known as the "pivot column" of this step of the simplex method; its entries are the numbers labelled α_i above. If no entry of the pivot column is positive, the algorithm terminates with the conclusion that f is not bounded above by the given constraints. Otherwise, for each positive entry of the pivot column, compute the entry of TB in the same row; these are the numbers labelled β_i above. The "pivot row" is the row among these in which the ratio β_i/α_i (which is nonnegative because α_i is positive and β_i is nonnegative) is as small as possible. Let k be the index of the pivot row. (Thus, if i is the index for which β_i/α_i is smallest, column i of TA is 1 in row k and 0 in all other rows.) Then, by definition, $T'A$ has the same column as TA in the $m-1$ columns corresponding to the basic variables other than x_i. In the pivot column, $T'A$ has column k of I_m. Let $P = T'T^{-1}$. If $\mathbf{e}$ is a column of I_m other than column k, then there is an integer r such that both TA and $T'A$ have $\mathbf{e}$ in column r. Therefore, $P\mathbf{e}$ is column r of $T'T^{-1}TA = T'A$, that is, $P\mathbf{e} = \mathbf{e}$. On the other hand, P times the pivot column is column k of I_m. These observations can be summarized by saying that $PQ = I$ where Q is the $m \times m$ matrix obtained from I_m by replacing column k with the pivot column. In summary:

Once the index of the pivot column is known, compute the entries of the pivot column of TA. Eliminate from consideration all rows in which the pivot column is not positive. (If no rows remain, f is not bounded above by the constraints.) For each of the remaining rows, compute the entry of TB in that row. The pivot row is the row in which the ratio of the entry in TB to the entry in the pivot column is smallest. (The next section gives a rule for breaking ties. In practice, ties are broken by making an arbitrary choice.) Let k be the index of the pivot row. Then $T' = Q^{-1}T$ where Q is the matrix obtained by replacing column k of I_m with the pivot column.

Each such step is called a "pivot." Since the algorithm must compute all the coefficients C_j at each step in order to determine the pivot column, instead of throwing them away and returning to the original expression of $f = c_1x_1 + c_2x_2 + \cdots + c_nx_n + d$ at each step, one should *update* the most recent expression $f = C_1x_1 + C_2x_2 + \cdots + C_nx_n + D$ in which $C_i = 0$ for all basic variables x_i. The algorithm for updating them can be described in the following way.

Let the pivot row, row k, of TA be computed. The entry of this row in column i is 1 and its entry in column j is α in the notation above. Subtract C_j/α times this row from the coefficients $C_1, C_2, \ldots, C_n$ to find the new coefficients C_1', $C_2', \ldots, C_n'$. (The updated value of D can be computed in a similar way, but this value is not needed.)

8 *Termination of the Simplex Method*

As was seen in Section 6, each pivot of the simplex algorithm replaces a feasible vertex with another feasible vertex which results in a *larger* value of f, except when the entry β of TB in the pivot row is zero. Therefore, as long as β is never zero, no feasible vertex is ever repeated. Since the number of feasible vertices is obviously finite (a vertex is a choice of basic variables, and there are at most $n!/(n-m)!$ choices), the algorithm must therefore reach a feasible vertex at which it terminates, either because the maximum has been found or because f has been shown not to be bounded above by the constraints.

The algorithm can always be carried out in such a way that it terminates, even when $\beta = 0$ occurs, as can be proved by proving that any problem can be *perturbed* slightly to obtain another problem for which the same pivots lead to a solution but for which $\beta = 0$ never occurs.

Specifically, the problem can be perturbed in the following way. Let a feasible vertex be given, and let the variables $x_1, x_2, \ldots, x_n$ of the problem be reordered, if necessary, to make the basic variables of the given feasible vertex $x_1, x_2, \ldots, x_m$ in that order. Perturb the problem by setting $x_i' = x_i + \epsilon^i$ for $i = 1, 2, \ldots, n$, where ϵ is a small positive rational number. Then $X' = X + E$ where E is an $n \times 1$ matrix whose entries are $\epsilon, \epsilon^2, \ldots, \epsilon^n$. The equations $TAX = TB$ of the original problem are replaced by the equations $TAX' = TAX + TAE = TB + TAE$ in the perturbed problem. In other words, TA is the same, but TB becomes $TB + TAE$. The entries of the $m \times 1$ matrix $TB + TAE$ are polynomials in ϵ in which the constant terms are the entries of TB (all of which are nonnegative) and in which the coefficient of the lowest power of ϵ is 1 because the first m columns of TA contain I_m. Therefore, the values of the m entries of $TB + TAE$ are all positive for all sufficiently small positive ϵ, which means that T makes explicit a feasible* vertex of the perturbed problem for all sufficiently small ϵ. Since the matrix TA is the same in the two problems, the choice of the pivot in the perturbed problem requires finding the *polynomial in* ϵ of the form $\beta_i(\epsilon)/\alpha_i$ that is smallest, where α_i is a positive entry in the pivot column of TA and $\beta_i(\epsilon)$ is the entry in the same row of $TB + TAE$. As long as there is not a tie in the choice of the pivot row of the unperturbed problem, the simplex method performs the same pivot in the two cases, because the polynomial with the smallest value for small ϵ is the one with the smallest constant term. If there is a tie in the choice of the pivot row in the unperturbed problem, the perturbed problem breaks the tie by looking at the terms containing ϵ. The choice is the same for all sufficiently small positive values of ϵ because a nonzero polynomial in ϵ has the same sign for all sufficiently small positive values of ϵ.

Subsequent pivots of the two problems proceed in the same way. At each step, there is an unambiguous choice for the pivot of the perturbed problem which is the same for all sufficiently small ϵ. (Let T'' make explicit the current feasible vertex. If two of the polynomials $\beta(\epsilon)/\alpha$ were equal, they would have to contain

---

*It was to ensure that the perturbation of the given feasible vertex would be feasible that the variables were reordered to put the basic variables first. What is needed precisely is that *for any zero entry of TB, the first nonzero entry of the corresponding row of TA should be positive.*

the same powers of ϵ, that is, two rows of $T''A$ would have to have nonzero entries in the same columns, which is contrary to the fact that all columns of I_m are contained in $T''A$.) This pivot is always a *possible* choice for the pivot of the unperturbed problem. For the perturbed problem, $\beta(\epsilon) = 0$ never occurs, so no vertex is ever repeated. Therefore, the perturbed problem gives a way of breaking ties in the choice of pivot row which guarantees that no vertex is ever repeated. Therefore, the simplex method, when carried out in this way, must terminate.

9 *Finding a Feasible Vertex*

The simplex method *assumes* that a feasible vertex of the given constraints is known. A standard trick of linear programming uses the simplex method itself to find a feasible vertex by setting up an *auxiliary* linear programming problem whose solution gives a feasible vertex.

Because a given set of constraints in standard form $AX = B$, $X \geq 0$ may be contradictory, the problem of finding a feasible vertex is better stated: Given an $m \times n$ matrix A of rank m and an $m \times 1$ matrix B, determine whether there is a feasible point for the constraints $AX = B$, $X \geq 0$, and, if so, find a feasible vertex.

Let a vertex of $AX = B$, $X \geq 0$ be given, let T make it explicit, and let s be the number of negative entries in TB. If $s = 0$, the vertex is feasible and nothing further is needed. Otherwise, let an auxiliary problem be constructed as follows.

To simplify the description of the auxiliary problem, let the m equations $TAX = TB$ be rearranged, if necessary, so that the s negative entries of TB appear in the first s equations, and then let the n variables $x_1, x_2, \ldots, x_n$ be rearranged, if necessary, so that the basic variables are $x_1, x_2, \ldots, x_m$, in that order. Then the equations $TAX = TB$ have the form $x_i + \alpha_{i,m+1}x_{m+1} + \alpha_{i,m+2}x_{m+2} + \cdots + \alpha_{i,n}x_n = \beta_i$ for $i = 1, 2, \ldots, m$, where $\beta_i < 0$ for $i \leq s$ and $\beta \geq 0$ for $i > s$. The auxiliary problem is: Subject to the constraints

$$\begin{aligned} u_i - x_i - \alpha_{i,m+1}x_{m+1} - \alpha_{i,m+2}x_{m+2} - \cdots - \alpha_{i,n}x_n &= -\beta_i \quad (i = 1, 2, \ldots, s) \\ x_i + \alpha_{i,m+1}x_{m+1} + \alpha_{i,m+2}x_{m+2} + \cdots + \alpha_{i,n}x_n &= \beta_i \quad (i = s+1, \ldots, m) \\ x_i \geq 0, \qquad u_i &\geq 0 \end{aligned} \tag{1}$$

on the $n + s$ variables $u_1, u_2, \ldots, u_s, x_1, x_2, \ldots, x_n$, find the maximum of $-u_1 - u_2 - \cdots - u_s$.

Because the right-hand sides of these equations are all nonnegative, the choice $u_1, u_2, \ldots, u_s, x_{s+1}, x_{s+2}, \ldots, x_m$ of basic variables gives a feasible vertex of the auxiliary problem. Therefore, the simplex method can be applied to the auxiliary problem. Because the constraints $u_i \geq 0$ obviously imply the upper bound 0 on $-u_1 - u_2 - \cdots - u_s$, it will terminate with a solution vertex of the auxiliary problem. In particular, the maximum of the auxiliary problem can be found.

If this maximum is less than 0, the original constraints must be contradictory, because a set of values for the xs that satisfied them would imply, when all us were set equal to zero, a set of values for the variables of the auxiliary problem for which the constraints were all satisfied and for which $-u_1 - u_2 - \cdots - u_s = 0$ was greater than the known maximum. What is to be shown, then, is that if the maximum is equal to 0 a feasible vertex of the original constraints can be found.

In case all the us are independent in the solution vertex of the auxiliary problem, so that all m of the basic variables of the auxiliary problem are xs, the choice of the same xs as basic variables of the original constraints gives a feasible vertex of the original constraints, as follows immediately from a consideration of what these conditions mean. Specifically, let A' and B' be the matrices that correspond to the constraints (1) of the auxiliary problem. In other words, let A' be the $m \times (n + s)$ matrix whose first s columns are the first s columns of I_m and whose remaining columns are JTA, where J is the matrix obtained by reversing the signs of the first s diagonal entries of I_m, and let $B' = JTB$. If S is the invertible $m \times m$ matrix that makes explicit the solution vertex of the auxiliary problem, then SA' is I_m in the columns corresponding to its basic variables and all entries of SB' are nonnegative. Therefore, SJT makes explicit the vertex of the original constraints in which the same xs are basic (because $SJTA$ is the identity matrix in the requisite columns) and all entries of $SJTB$ are nonnegative, which shows that this is a feasible vertex of the original problem.

When one or more of the us are basic variables in the solution vertex of the auxiliary problem, one can perform pivots to obtain another solution vertex in which fewer us are basic in the following way. As above, let S make explicit a solution vertex of the auxiliary problem. If u_i is a basic variable of this vertex, then column i of SA' is a column of I_m, say it is column k, that is, its 1 is in the kth row. There must be a nonzero entry in each row of the n columns of SA' corresponding to the xs, because otherwise $SJTA$ would have a zero row, contrary to the facts that S, J, and T are all invertible and A has rank m. In particular, row k of SA' must have a nonzero entry in the column corresponding to one of the xs, say in the column corresponding to x_j. Then replacement of u_i by x_j as a basic variable gives a vertex of the auxiliary problem, and this vertex is made explicit by the matrix $S' = Q^{-1}S$, where Q is the matrix obtained by replacing column k of I_m with the column of SA' corresponding to x_j. (See Section 7. $P = S'S^{-1}$ times any column of Q is the corresponding column of I_m, that is, $PQ = I_m$.) It is to be shown that this vertex of the auxiliary problem is a solution vertex.

Setting the independent variables of a vertex equal to zero gives the basic variables the values in the corresponding rows of SB', where S makes the vertex explicit. Since all us must be zero at a point where $-u_1 - u_2 - \cdots - u_s$ assumes its maximum, it follows that the kth entry of SB' is zero. Thus, SB' is a sum of multiples of columns of I_m other than column k. Since multiplication on the left by Q^{-1} carries any one of these columns to itself, $Q^{-1}SB' = SB'$, that is, $S'B' = SB'$. Therefore, for each of the $m - 1$ variables that are basic in both vertices (those corresponding to rows other than the kth), the values that result when all independent variables are set equal to zero are the same for the two vertices. Since all other variables have the value zero in both cases, all variables have the same values in both. In particular, since all us have the value zero in the first case, the same is true in the other case, which means that the new vertex is a solution vertex of the auxiliary problem.

Repetition of this step for as long as some u is basic produces finally a solution vertex of the auxiliary problem in which all us are independent. Therefore, as was shown above, it produces a feasible vertex of the original constraints, as was to be shown.

10 Summary

The constraints of an arbitrary linear programming problem can be stated in standard form or they can be shown to be contradictory (Section 2). There is a vertex for constraints in this form (Section 3). If a given vertex is not feasible, the auxiliary problem described in Section 9 is a linear programming problem for which a feasible vertex is known. The simplex algorithm can then be applied (Section 6, Section 7) to the auxiliary problem to find a sequence of pivots leading to new feasible vertices and terminating (Section 8) with a solution vertex of the auxiliary problem. If the maximum of the auxiliary problem is not 0, the original constraints were contradictory. Otherwise, a solution vertex of the auxiliary problem can be found in which the basic variables are all variables of the original constraints, and such a solution vertex yields a feasible vertex of the original constraints (Section 9).

Once a feasible vertex of the original problem has been found, the simplex algorithm can be applied to the given problem. The result is either a solution vertex or a restatement of the problem in a form showing that the given constraints imply no upper bound on the function to be maximized.

Examples

As a first example, consider the problem of maximizing $4x + 3y$ subject to

$$\begin{aligned} 2x + y &\le 25 \\ -3x + 2y &\le 15 \\ x + y &\le 15. \end{aligned}$$

To make these equality constraints with the statement that some variables are special, write them as

$$\begin{gathered} 2x + y + u = 25 \\ -3x + 2y + v = 15 \\ x + y + w = 15 \\ u \ge 0, \quad v \ge 0, \quad w \ge 0. \end{gathered}$$

The first equation can be used to write $y = 25 - 2x - u$ and to eliminate y from the other two equations. The last equation becomes $-x - u + w = -10$, so $x = -u + w + 10$ can be used to eliminate x from the middle equation and to eliminate x from the expression of y. The result is

$$\begin{gathered} -u \quad + w + 10 = x \\ u \quad - 2w + 5 = y \\ 5u + v - 7w = 35 \\ u \ge 0, \quad v \ge 0, \quad w \ge 0. \end{gathered}$$

Subject to the single constraint $5u + v - 7w = 35$, the maximum of $4x + 3y = -u - 2w + 55$ is to be found. Since $u, w \ge 0$, this maximum is obviously 55, which occurs only where $u = w = 0$, at which point $v = 35$, $x = 10$, and $y = 5$.

Consider next the problem of maximizing $x+y+2z$ subject to the constraints

$$\begin{aligned} 2x + y &\ge 2 \\ x - 2y &\le 4 \\ 3x + z &= 5 \\ y + z &\le 0 \\ -x + y + 3z &\le 8. \end{aligned}$$

The procedure of Section 2 calls for replacing the first constraint with the equation $-2x - y + u = -2$ and the inequality $u \ge 0$. The second constraint becomes, in the same way, $x - 2y + v = 4$ and $v \ge 0$, the fourth constraint $y + z + w = 0$,

$w \geq 0$, and the fifth constraint $-x+y+3z+t=8$, $t \geq 0$. Thus, the constraints are

$$\begin{aligned} -2x - y + u &= -2 \\ x - 2y + v &= 4 \\ 3x + z &= 5 \\ y + z + w &= 0 \\ -x + y + 3z + t &= 8 \\ u \geq 0, \quad v \geq 0, \quad w &\geq 0, \quad t \geq 0. \end{aligned}$$

The nonspecial variables x, y, and z can be eliminated as follows. By the third equation, $z = 5 - 3x$ and z can be eliminated from the other equations to find

$$\begin{aligned} -2x - y + u &= -2 \\ x - 2y + v &= 4 \\ -3x + y + w &= -5 \\ -10x + y + t &= -7 \\ u \geq 0, \quad v \geq 0, \quad w &\geq 0, \quad t \geq 0. \end{aligned}$$

By the first equation $y = -2x + u + 2$ and y can be eliminated from the other equations to find

$$\begin{aligned} 5x - 2u + v &= 8 \\ -5x + u + w &= -7 \\ -12x + u + t &= -9 \\ u \geq 0, \quad v \geq 0, \quad w &\geq 0, \quad t \geq 0. \end{aligned}$$

Finally, by the first equation $x = \frac{2}{5}u - \frac{1}{5}v + \frac{8}{5}$ and the equations relating the special variables can be expressed in the form

$$\begin{aligned} -u + v + w &= 1 \\ -19u + 12v + 5t &= 51 \\ u \geq 0, \quad v \geq 0, \quad w &\geq 0, \quad t \geq 0. \end{aligned}$$

These constraints are in standard form $AX = B$, $X \geq 0$, where X is the 4×1 matrix containing the variables u, v, w, t in that order, where $A = \begin{bmatrix} -1 & 1 & 1 & 0 \\ -19 & 12 & 0 & 5 \end{bmatrix}$ has rank 2, and $B = \begin{bmatrix} 1 \\ 51 \end{bmatrix}$. The list (3, 4) of indices gives a vertex of these constraints which is made explicit by the matrix $\begin{bmatrix} 1 & 0 \\ 0 & \frac{1}{5} \end{bmatrix}$. When the independent variables u, v of this vertex are set equal to zero, the values of the basic variables are given by $TB = \begin{bmatrix} 1 & 0 \\ 0 & \frac{1}{5} \end{bmatrix}\begin{bmatrix} 1 \\ 51 \end{bmatrix} = \begin{bmatrix} 1 \\ \frac{51}{5} \end{bmatrix}$. Since these values are nonnegative, the vertex is *feasible* and the simplex method can be applied.

The function $x + y + 2z$ to be maximized can be expressed in terms of the independent variables u and v by means of the substitutions $x + y + 2z = x+y+2(5-3x) = -5x+y+10 = -5x+(-2x+u+2)+10 = -7x+u+12 = -\frac{7}{5}(2u - v + 8) + u + 12 = -\frac{9}{5}u + \frac{7}{5}v + \frac{4}{5}$. (It happens more or less by chance that w and t do not appear in this function, so the equations $AX = B$ do not need to be used to eliminate them.) The positive coefficient of v in this function implies that the simplex method calls for a pivot in which the v column is the pivot column. The entries of this column of TA are 1 and $\frac{12}{5}$, both of which are positive. The entries of TB, as already noted, are 1, and $\frac{51}{5}$. The pivot row is determined by comparing the ratios $1/1$ and $\frac{51}{5}/\frac{12}{5} = \frac{51}{12}$. Since the first one is smaller, the pivot row is the first row; therefore, the next vertex is described by the list (2, 4) of basic variables, and it is made explicit by the matrix

$$\begin{bmatrix} 1 & 0 \\ \frac{12}{5} & 1 \end{bmatrix}^{-1}\begin{bmatrix} 1 & 0 \\ 0 & \frac{1}{5} \end{bmatrix} = \begin{bmatrix} 1 & 0 \\ -\frac{12}{5} & \frac{1}{5} \end{bmatrix}.$$

(Multiplication on the left by the inverse of $\begin{bmatrix} 1 & 0 \\ \frac{12}{5} & 1 \end{bmatrix}$ subtracts $\frac{12}{5}$ times the first row from the second.) The coefficients of the function to be maximized were $-\frac{9}{7}$, $\frac{7}{5}$, 0, 0 and the pivot row was -1, 1, 1, 0 so the new coefficients of the function to be maximized are $-\frac{2}{5}$, 0, $-\frac{7}{5}$, 0. Since no coefficient is positive, this vertex is a solution vertex. In fact, the coefficients of the independent variables are both negative, so the maximum is achieved only when the two independent variables of the vertex are zero. The corresponding values of the basic variables v and t are the entries of $\begin{bmatrix} 1 & 0 \\ -\frac{12}{5} & \frac{1}{5} \end{bmatrix}\begin{bmatrix} 1 \\ 51 \end{bmatrix}$, so they are $v = 1$ and $t = -\frac{39}{5}$. The desired maximum therefore occurs when $x = \frac{1}{5}(2u - v + 8) = \frac{7}{5}$, $z = 5 - 3x = \frac{4}{5}$, $y = -2x + u + 2 = -\frac{4}{5}$, and the maximum is $x + y + 2z = \frac{11}{5}$.

As a final example, consider the problem of minimizing $x + 4y$ subject to the constraints

$$\begin{aligned} x + y &\geq 1 \\ 2x + 3y &\geq 2 \\ x - y &\geq -1 \\ x \geq 0,\ y &\geq 0. \end{aligned}$$

When the constraints are put in standard form they become

$$-x - y + a = -1$$
$$-2x - 3y + b = -2$$
$$-x + y + c = 1$$

where all variables are special. In matrix form, $AX = B$ and $X \geq 0$, where

$$A = \begin{bmatrix} -1 & -1 & 1 & 0 & 0 \\ -2 & -3 & 0 & 1 & 0 \\ -1 & 1 & 0 & 0 & 1 \end{bmatrix}, \qquad B = \begin{bmatrix} -1 \\ -2 \\ 1 \end{bmatrix}$$

and X is the 5×1 matrix whose entries are the variables x, y, a, b, c.

Here a, b, c are the basic variables of a vertex that is made explicit by I_3. However, because the entries of $I_3B = B$ are not positive, this vertex is not feasible. The construction of Section 9 leads to the auxiliary problem $A'U = B'$, $U \geq 0$ where

$$A' = \begin{bmatrix} 1 & 0 & -1 & 0 & 0 & 1 & 1 \\ 0 & 1 & 0 & -1 & 0 & 2 & 3 \\ 0 & 0 & 0 & 0 & 1 & -1 & 1 \end{bmatrix} \qquad B' = \begin{bmatrix} 1 \\ 2 \\ 1 \end{bmatrix}$$

and where U is the 7×1 matrix whose entries are the variables u, v, a, b, c, x, y. (The order of the variables is more or less arbitrary.) The objective of the auxiliary problem is to find the minimum of $-u - v$. The vertex with basic variables u, v, c is made explicit by the matrix I_3. This vertex is feasible because the entries of B' are positive. The expression of $-u - v$ in terms of the independent variables of this vertex is $-(a - x - y + 1) - (b - 2x - 3y + 2) = -a - b + 3x + 4y - 3$, so application of the simplex method to this vertex of the auxiliary problem calls for a pivot in which the y column is the pivot column.

The entries 1, 3, 1 of the pivot column are all positive. The pivot row is the second row, because the ratios that determine the pivot row are $1/1$, $2/3$, and $1/1$. Therefore, the basic variables of the next vertex are u, y, c and it is made explicit by

$$S = \begin{bmatrix} 1 & 1 & 0 \\ 0 & 3 & 0 \\ 0 & 1 & 1 \end{bmatrix}^{-1} = \begin{bmatrix} 1 & -\frac{1}{3} & 0 \\ 0 & \frac{1}{3} & 0 \\ 0 & -\frac{1}{3} & 1 \end{bmatrix}.$$

The coefficients of the objective function were 0, 0, -1, -1, 0, 3, 4 and the entries of the pivot row were 0, 1, 0, -1, 0, 2, 3, so the new coefficients are 0, $-\frac{4}{3}$, -1, $\frac{1}{3}$, 0, $\frac{1}{3}$, 0. Another pivot is called for, and the choice of pivot column is tied between b and x. Let b be chosen.

The pivot column is

$$S \begin{bmatrix} 0 \\ -1 \\ 0 \end{bmatrix} = \begin{bmatrix} \frac{1}{3} \\ -\frac{1}{3} \\ \frac{1}{3} \end{bmatrix}.$$

The first and third entries of SB are both $\frac{1}{3}$, so the choice of pivot row is tied between the first and third rows, because in both cases the ratio is $\frac{1}{3}/\frac{1}{3} = 1$. Let the third row be chosen as the pivot row. The basic variables become u, y, b and the matrix that makes the new vertex explicit is

$$S' = \begin{bmatrix} 1 & 0 & \frac{1}{3} \\ 0 & 1 & -\frac{1}{3} \\ 0 & 0 & \frac{1}{3} \end{bmatrix}^{-1} \begin{bmatrix} 1 & -\frac{1}{3} & 0 \\ 0 & \frac{1}{3} & 0 \\ 0 & -\frac{1}{3} & 1 \end{bmatrix} = \begin{bmatrix} 1 & 0 & -1 \\ 0 & 0 & 1 \\ 0 & -1 & 3 \end{bmatrix}.$$

(Multiplication on the left by this inverse is the same as subtraction of the last row from the first, addition of the last row to the second, and multiplication of the last row by 3.) The coefficients of the objective function were $0, -\frac{4}{3}, -1, \frac{1}{3}, 0, \frac{1}{3}, 0$ and the entries of the pivot row were the entries of the last row of SA', which can easily be found to be $0, -\frac{1}{3}, 0, \frac{1}{3}, 1, -\frac{5}{3}, 0$ from which it follows that the new coefficients are $0, -1, -1, 0, -1, 2, 0$. Another pivot is called for, and the pivot column is the column of x.

The entries of the pivot column are

$$S' \begin{bmatrix} 1 \\ 2 \\ -1 \end{bmatrix} = \begin{bmatrix} 2 \\ -1 \\ -5 \end{bmatrix}$$

which shows that the first row is the pivot row. The basic variables of the new vertex are x, y, b and it is made explicit by

$$S'' = \begin{bmatrix} 2 & 0 & 0 \\ -1 & 1 & 0 \\ -5 & 0 & 1 \end{bmatrix}^{-1} \begin{bmatrix} 1 & 0 & -1 \\ 0 & 0 & 1 \\ 0 & -1 & 3 \end{bmatrix} = \begin{bmatrix} \frac{1}{2} & 0 & -\frac{1}{2} \\ \frac{1}{2} & 0 & \frac{1}{2} \\ \frac{5}{2} & -1 & \frac{1}{2} \end{bmatrix}.$$

Since both u and v are now independent, the expression of the objective function of the auxiliary problem in terms of the independent variables is simply $-u - v$ and it assumes its maximum value 0 when $u = v = 0$.

Therefore, x, y, b are the basic variables of a feasible vertex of the *original* problem, and this vertex is made explicit by the matrix

$$T' = S''JT = \begin{bmatrix} \frac{1}{2} & 0 & -\frac{1}{2} \\ \frac{1}{2} & 0 & \frac{1}{2} \\ \frac{5}{2} & -1 & \frac{1}{2} \end{bmatrix} \begin{bmatrix} -1 & 0 & 0 \\ 0 & -1 & 0 \\ 0 & 0 & 1 \end{bmatrix} I_3 = \begin{bmatrix} -\frac{1}{2} & 0 & -\frac{1}{2} \\ -\frac{1}{2} & 0 & \frac{1}{2} \\ -\frac{5}{2} & 1 & \frac{1}{2} \end{bmatrix}.$$

The objective of the problem is to minimize $x + 4y$, which is to maximize $-x - 4y$. Computation of $T'A$ and $T'B$ gives

$$T'A = \begin{bmatrix} 1 & 0 & -\frac{1}{2} & 0 & -\frac{1}{2} \\ 0 & 1 & -\frac{1}{2} & 0 & \frac{1}{2} \\ 0 & 0 & -\frac{5}{2} & 1 & \frac{1}{2} \end{bmatrix}, \qquad T'B = \begin{bmatrix} 0 \\ 1 \\ 1 \end{bmatrix}.$$

Therefore $x - \frac{1}{2}a - \frac{1}{2}c = 0$ and $y - \frac{1}{2}a + \frac{1}{2}c = 1$, so $-x - 4y = -(\frac{1}{2}a + \frac{1}{2}c) - 4(\frac{1}{2}a - \frac{1}{2}c + 1) = -\frac{5}{2}a + \frac{3}{2}c - 4$. The simplex method calls for another pivot in which the pivot column is the column of c.

The entries of the pivot column are $-\frac{1}{2}, \frac{1}{2}, \frac{1}{2}$ and the last two entries of $T'B$ are both 1, so there is a tie between the last two rows for the choice of pivot row. Let the last row be chosen. Then the new basic variables are x, y, c and the new vertex is made explicit by

$$T'' = \begin{bmatrix} 1 & 0 & -\frac{1}{2} \\ 0 & 1 & \frac{1}{2} \\ 0 & 0 & \frac{1}{2} \end{bmatrix}^{-1} \begin{bmatrix} -\frac{1}{2} & 0 & -\frac{1}{2} \\ -\frac{1}{2} & 0 & \frac{1}{2} \\ -\frac{5}{2} & 1 & \frac{1}{2} \end{bmatrix} = \begin{bmatrix} -3 & 1 & 0 \\ 2 & -1 & 0 \\ -5 & 2 & 1 \end{bmatrix}.$$

The coefficients of the objective function were $0, 0, -\frac{5}{2}, 0, \frac{3}{2}$ and the pivot row was $0, 0, -\frac{5}{2}, 1, \frac{1}{2}$, so the new coefficients are $0, 0, 5, -3, 0$. The simplex method calls for another pivot in which the pivot column is the column of a.

The pivot column is

$$T'' \begin{bmatrix} 1 \\ 0 \\ 0 \end{bmatrix} = \begin{bmatrix} -3 \\ 2 \\ -5 \end{bmatrix}$$

so the pivot row is the second row. The new basic variables are x, a, c and the vertex is made explicit by

$$T^{(3)} = \begin{bmatrix} 1 & -3 & 0 \\ 0 & 2 & 0 \\ 0 & -5 & 1 \end{bmatrix}^{-1} \begin{bmatrix} -3 & 1 & 0 \\ 2 & -1 & 0 \\ -5 & 2 & 1 \end{bmatrix} = \begin{bmatrix} 0 & -\frac{1}{2} & 0 \\ 1 & -\frac{1}{2} & 0 \\ 0 & -\frac{1}{2} & 1 \end{bmatrix}.$$

The coefficients of the objective function were $0, 0, 5, -3, 0$ and the entries of the pivot row were $0, \frac{1}{2}, 1, -\frac{1}{2}, 0$ (the middle row of $T''A$), so the new coefficients are $0, -\frac{5}{2}, 0, -\frac{1}{2}, 0$. Since the coefficients of both independent variables are negative, the maximum value is attained only when these variables are both zero. The values of the basic variables are the entries of $T^{(3)}B$, which are 1, 0, 2, in that order. Therefore, the maximum of $-x - 4y$ occurs only when $x = 1, a = 0, c = 2, y = 0, b = 0$, and the function $x + 4y$ attains its minimum value subject to the given constraints only when $x = 1$ and $y = 0$. This minimum value is of course 1.

Answers to Exercises

Chapter 1

1

$$u = 3(a + b + c) + 2(3a - 2b + 4c) = 9a - b + 11c$$

$$v = 7(a + b + c) - 5(3a - 2b + 4c) = -8a + 17b - 13c$$

$$w = -(a + b + c) + 11(3a - 2b + 4c) = 32a - 23b + 43c$$

or

$$\begin{bmatrix} 3 & 2 \\ 7 & -5 \\ -1 & 11 \end{bmatrix} \begin{bmatrix} 1 & 1 & 1 \\ 3 & -2 & 4 \end{bmatrix} = \begin{bmatrix} 9 & -1 & 11 \\ -8 & 17 & -13 \\ 32 & -23 & 43 \end{bmatrix}$$

2

$$\begin{bmatrix} 1 & 1 & 1 \\ 3 & -2 & 4 \end{bmatrix} \begin{bmatrix} 3 & 2 \\ 7 & -5 \\ -1 & 11 \end{bmatrix} = \begin{bmatrix} 9 & 8 \\ -9 & 60 \end{bmatrix}.$$

Note that the matrix product is of the same two matrices as in Exercise 1 but in the opposite order. Comparison of the two answers shows that matrix multiplication does not commute.

3 The composed function is

$$p = 33a - 7b + 41c$$

$$q = 171a - 129b + 231c.$$

In terms of matrix multiplication, this exercise illustrates the associative law of matrix multiplication. The triple product

$$\begin{bmatrix} 1 & 1 & 1 \\ 3 & -2 & 4 \end{bmatrix} \begin{bmatrix} 3 & 2 \\ 7 & -5 \\ -1 & 11 \end{bmatrix} \begin{bmatrix} 1 & 1 & 1 \\ 3 & -2 & 4 \end{bmatrix}$$

can be computed either as

$$= \begin{bmatrix} 1 & 1 & 1 \\ 3 & -2 & 4 \end{bmatrix} \begin{bmatrix} 9 & -1 & 11 \\ -8 & 17 & -13 \\ 32 & -23 & 43 \end{bmatrix} = \begin{bmatrix} 33 & -7 & 41 \\ 171 & -129 & 231 \end{bmatrix}$$

or as

$$= \begin{bmatrix} 9 & 8 \\ -9 & 60 \end{bmatrix} \begin{bmatrix} 1 & 1 & 1 \\ 3 & -2 & 4 \end{bmatrix} = \begin{bmatrix} 33 & -7 & 41 \\ 171 & -129 & 231 \end{bmatrix}.$$

4

$$u = 10a + 4b + c$$

$$v = 2a + 4b - 3c.$$

5 Direct computation shows that the square of J is

$$\begin{bmatrix} 0 & -1 \\ 1 & 0 \end{bmatrix}\begin{bmatrix} 0 & -1 \\ 1 & 0 \end{bmatrix} = \begin{bmatrix} -1 & 0 \\ 0 & -1 \end{bmatrix}$$

and that the square of this matrix is $\begin{bmatrix} 1 & 0 \\ 0 & 1 \end{bmatrix}$. The matrix $\begin{bmatrix} 1 & 0 \\ 0 & 1 \end{bmatrix}$ is called the 2×2 *identity matrix* and is denoted I_2. It has the property, as is easily checked, that $IM = M$ whenever M is a matrix with two rows and $MI = M$ whenever M is a matrix with two columns. Thus, if $n > 4$ we have $J^n = J^4 J^{n-4} = IJ^{n-4} = J^{n-4}$, as was to be shown. Note that J describes the linear substitution $x = -y$, $y = x$, which can be pictured as a rotation by 90° degrees. With this interpretation, it becomes clear that four iterations of the substitution produce the identity.

6 (a) The first few cases $A^2 = \begin{bmatrix} 1 & 2 \\ 0 & 1 \end{bmatrix}$, $A^3 = \begin{bmatrix} 1 & 3 \\ 0 & 1 \end{bmatrix}$, $A^4 = \begin{bmatrix} 1 & 4 \\ 0 & 1 \end{bmatrix}$ lead to the guess $A^n = \begin{bmatrix} 1 & n \\ 0 & 1 \end{bmatrix}$. If this formula is correct for n it is correct for $n+1$, because then $A^{n+1} = AA^n = \begin{bmatrix} 1 & 1 \\ 0 & 1 \end{bmatrix}\begin{bmatrix} 1 & n \\ 0 & 1 \end{bmatrix} = \begin{bmatrix} 1 & n+1 \\ 0 & 1 \end{bmatrix}$. Therefore, it is correct for all positive integers. (b) A is the matrix of coefficients of the linear substitution $x = x' + y'$, $y = y'$. The inverse of this substitution expresses x' and y' in terms of x and y. It can be found simply by subtracting the second equation from the first to find $x - y = x'$, $y = y'$. The matrix of coefficients of this substitution is $\begin{bmatrix} 1 & -1 \\ 0 & 1 \end{bmatrix}$. (c) The answer to (b) can be regarded as the case $n = -1$ of the formula of (a).

7 Multiplication of B on the left by A acts on each column of B in the same way. Therefore, it suffices to consider the case in which B has only one column. If the entries of a 3×1 matrix B are indeterminates, say $B = \begin{bmatrix} a \\ b \\ c \end{bmatrix}$, then $AB = \begin{bmatrix} a \\ b+c \\ c \end{bmatrix}$, which shows clearly that the effect of multiplication on the left by A is to add the last row to the middle one, without changing the first or last row. Similarly, the effect of multiplication on the right by A is to add the middle column to the last without changing the first two columns.

8 (a) $A = \begin{bmatrix} 1 & 0 & 0 & 0 \\ 0 & 1 & 0 & 0 \\ 0 & 1 & 1 & 0 \\ 0 & 0 & 0 & 1 \end{bmatrix}$. (b) $A = \begin{bmatrix} 1 & 0 & 0 & 0 \\ 0 & 1 & 1 & 0 \\ 0 & 0 & 1 & 0 \\ 0 & 0 & 0 & 1 \end{bmatrix}$.

9 The inverse operation of adding the second row to the third is the operation of subtracting the second row from the third, in the sense that doing first one of these operations then the other—in either order—results in no change. This operation is the same as multiplication on the left by $\begin{bmatrix} 1 & 0 & 0 & 0 \\ 0 & 1 & 0 & 0 \\ 0 & -1 & 1 & 0 \\ 0 & 0 & 0 & 1 \end{bmatrix}$, so this is the inverse of the matrix in part (a). Similarly, the inverse of the matrix in part (b) is $\begin{bmatrix} 1 & 0 & 0 & 0 \\ 0 & 1 & -1 & 0 \\ 0 & 0 & 1 & 0 \\ 0 & 0 & 0 & 1 \end{bmatrix}$.

10

$$AB = \begin{bmatrix} 3 \\ 6 \end{bmatrix}, \quad AC = \begin{bmatrix} 4 & 5 & 6 \\ 8 & 10 & 12 \end{bmatrix},$$

$$BC = \begin{bmatrix} 12 & 15 & 18 \end{bmatrix}, \quad B^2 = \begin{bmatrix} 9 \end{bmatrix}, \quad CE = \begin{bmatrix} 32 \end{bmatrix},$$

$$DA = \begin{bmatrix} 2 \\ -1 \end{bmatrix}, \quad D^2 = \begin{bmatrix} -1 & 0 \\ 0 & -1 \end{bmatrix}, \quad EB = \begin{bmatrix} 3 \\ 6 \\ 9 \end{bmatrix}, \quad EC = \begin{bmatrix} 4 & 5 & 6 \\ 8 & 10 & 12 \\ 12 & 15 & 18 \end{bmatrix}.$$

11 Self-checking.

12

$$BAC = \begin{bmatrix} 6 & 10 \end{bmatrix}$$

Chapter 2

1

(a) $$\begin{bmatrix} 1 & 0 \\ -1 & 1 \end{bmatrix}\begin{bmatrix} 4 & 2 \\ -2 & -4 \end{bmatrix}\begin{bmatrix} 1 & -1 \\ -1 & 2 \end{bmatrix} = \begin{bmatrix} 2 & 0 \\ 0 & -6 \end{bmatrix}$$

(b) $$\begin{bmatrix} 1 & 0 & 0 \\ 3 & -3 & -1 \\ -3 & 4 & 1 \end{bmatrix}\begin{bmatrix} 2 & 1 \\ 1 & 1 \\ 2 & -1 \end{bmatrix}\begin{bmatrix} 1 & -1 \\ -1 & 2 \end{bmatrix} = \begin{bmatrix} 1 & 0 \\ 0 & 1 \\ 0 & 0 \end{bmatrix}$$

(c) $$\begin{bmatrix} 1 & -8 & 6 \\ -1 & 12 & -9 \\ 0 & -5 & 4 \end{bmatrix}\begin{bmatrix} 3 \\ 4 \\ 5 \end{bmatrix} = \begin{bmatrix} 1 \\ 0 \\ 0 \end{bmatrix}$$

(d) $$\begin{bmatrix} 1 & 0 \\ -3 & 1 \end{bmatrix}\begin{bmatrix} 3 & 2 & 2 \\ 5 & 2 & 1 \end{bmatrix}\begin{bmatrix} 1 & 2 & -2 \\ -1 & -6 & 7 \\ 0 & 3 & -4 \end{bmatrix} = \begin{bmatrix} 1 & 0 & 0 \\ 0 & 1 & 0 \end{bmatrix}$$

2 Self-checking.

3 T commutes with itself and T^{-1}, but it does not commute with the other two 2×2 tilts. The four 3×3 tilts $\begin{bmatrix} 1 & \pm 1 & 0 \\ 0 & 1 & 0 \\ 0 & 0 & 1 \end{bmatrix}$ and $\begin{bmatrix} 1 & 0 & 0 \\ 0 & 1 & 0 \\ 0 & \pm 1 & 1 \end{bmatrix}$ fail to commute with the given tilt; the other four do commute with it.

4 $4n - 4$.

5 A tilt has two nonzero entries in exactly one column and two nonzero entries in exactly one row. Two tilts T_1 and T_2 commute unless the number of the column in which T_1 has two nonzero entries is the same as the number of the row in which T_2 has two nonzero entries or *vice versa*. When the tilts are viewed as row operations, the rule is that one row addition or subtraction commutes with another unless the row being added or subtracted by one is the row to which a row is added or subtracted by the other.

6 Since every matrix is equivalent to a diagonal matrix, $\begin{bmatrix} a & b \end{bmatrix}$ is equivalent to a matrix of the form $\begin{bmatrix} c & 0 \end{bmatrix}$. The equivalences $\begin{bmatrix} c & 0 \end{bmatrix} \sim \begin{bmatrix} c & -c \end{bmatrix} \sim \begin{bmatrix} 0 & -c \end{bmatrix} \sim \begin{bmatrix} -c & -c \end{bmatrix} \sim \begin{bmatrix} -c & 0 \end{bmatrix}$ show that every diagonal 1×2 matrix is equivalent to one in which the diagonal entry is nonnegative. It remains only to show that if $\begin{bmatrix} d & 0 \end{bmatrix} \sim \begin{bmatrix} e & 0 \end{bmatrix}$ and $d \geq 0$ and $e \geq 0$, then $d = e$. Since both entries of $\begin{bmatrix} d & 0 \end{bmatrix}$ are multiples of d, so are both entries of any matrix equivalent to $\begin{bmatrix} d & 0 \end{bmatrix}$, because a sum or difference of multiples of d is a multiple of d. Therefore, e is a multiple of d, say $e = id$. By the same token, d is a multiple of e, say $d = je$. If $d = 0$ then $e = id = 0$. Otherwise, division of $d = je = ijd$ by d gives $1 = ij$. Since i and j are integers, it follows that $i = \pm 1$. Since $d > 0$ and $e \geq 0$, i cannot be -1. Therefore, $d = e$, as was to be shown.

7 The algorithm of Section 6 applied to $\begin{bmatrix} a & b \end{bmatrix}$ produces a sequence of linked matrices $\begin{bmatrix} a & b \end{bmatrix} \sim \cdots \sim \begin{bmatrix} d & 0 \end{bmatrix}$ in which all entries except the last entry of the last matrix are positive. Each matrix in this sequence is equal to the following matrix multiplied on the right by R or L. Thus $\begin{bmatrix} a & b \end{bmatrix}$ is equal to $\begin{bmatrix} d & 0 \end{bmatrix}$ multiplied on the right by a sequence of Rs and/or Ls.

8 (a) $\begin{bmatrix} 9 & 15 \end{bmatrix} = \begin{bmatrix} 3 & 0 \end{bmatrix} R^2LR = \begin{bmatrix} 3 & 0 \end{bmatrix} \begin{bmatrix} 3 & 5 \\ 1 & 2 \end{bmatrix}$.

(b) $\begin{bmatrix} 12 & 17 \end{bmatrix} = \begin{bmatrix} 1 & 0 \end{bmatrix} RLR^2L^2R = \begin{bmatrix} 1 & 0 \end{bmatrix} \begin{bmatrix} 12 & 17 \\ 7 & 10 \end{bmatrix}$.

(c) $\begin{bmatrix} 37 & 27 \end{bmatrix} = \begin{bmatrix} 1 & 0 \end{bmatrix} RL^2R^2LR^2L = \begin{bmatrix} 1 & 0 \end{bmatrix} \begin{bmatrix} 37 & 27 \\ 26 & 19 \end{bmatrix}$.

(d) $\begin{bmatrix} 104 & 65 \end{bmatrix} = \begin{bmatrix} 13 & 0 \end{bmatrix} R^2LRL = \begin{bmatrix} 13 & 0 \end{bmatrix} \begin{bmatrix} 8 & 5 \\ 3 & 2 \end{bmatrix}$.

(e) $\begin{bmatrix} 42 & 53 \end{bmatrix} = \begin{bmatrix} 1 & 0 \end{bmatrix} R^2L^4RL^3R = \begin{bmatrix} 1 & 0 \end{bmatrix} \begin{bmatrix} 42 & 53 \\ 19 & 24 \end{bmatrix}$.

(f) $\begin{bmatrix} 54 & 64 \end{bmatrix} = \begin{bmatrix} 2 & 0 \end{bmatrix} RLR^2L^5R = \begin{bmatrix} 2 & 0 \end{bmatrix} \begin{bmatrix} 27 & 32 \\ 16 & 19 \end{bmatrix}$.

9 If N is a product of Rs and Ls, if $\begin{bmatrix} r & s \end{bmatrix}$ is the first row of N, and if the rightmost factor of N is L, then $r > s$; if the rightmost factor is R, then $r \le s$. Therefore, the rightmost factors in two equal products must be the same. Multiplication on the right by the inverse of this common rightmost factor gives two shorter products of R and L that are equal. Repetition of this process reduces the statement to be proved to the statement that no product of Rs and Ls is equal to I. If the leftmost factor of a product is L, both entries of the top row of the product are positive, and therefore the product is not I. Similarly, if the leftmost factor is R then both entries of the bottom row of the product are positive and the product is not I, which completes the proof.

10 If N is a product of *no* factors, that is, if $N = I$, then $ru = 1 \cdot 1 = 0 \cdot 0 + 1 = st + 1$. If $\begin{bmatrix} r & s \\ t & u \end{bmatrix} = \begin{bmatrix} r' & s' \\ t' & u' \end{bmatrix} \begin{bmatrix} 1 & 0 \\ 1 & 1 \end{bmatrix}$ and if r', s', t', u' are nonnegative integers for which $r'u' = s't' + 1$, then $r = r' + s'$, $s = s'$, $t = t' + u'$, $u = u'$ are nonnegative integers; $ru = (r' + s')u' = r'u' + s'u'$; and $st + 1 = s'(t' + u') + 1 = s't' + 1 + s'u' = r'u' + s'u' = ru$. In the same way, $\begin{bmatrix} r & s \\ t & u \end{bmatrix} = \begin{bmatrix} r' & s' \\ t' & u' \end{bmatrix} \begin{bmatrix} 1 & 1 \\ 0 & 1 \end{bmatrix}$ and $r'u' = s't' + 1$ imply $ru = st + 1$.

11 By the preceding exercises, $\begin{bmatrix} a & b \end{bmatrix} = \begin{bmatrix} d & 0 \end{bmatrix} \begin{bmatrix} r & s \\ t & u \end{bmatrix}$ where $ru = st + 1$. Thus, $a = dr$, $b = ds$, $ua = dru = d(st + 1) = dst + d = tb + d$, so the desired equation holds with $p = u$ and $q = t$.

12 Because $\begin{bmatrix} 59 & 26 \end{bmatrix} = \begin{bmatrix} 1 & 0 \end{bmatrix} RLR^2LR^3L^2$ and $RLR^2LR^3L^2 = \begin{bmatrix} 59 & 26 \\ 34 & 15 \end{bmatrix}$, the desired equation is $15 \cdot 59 = 885 = 34 \cdot 26 + 1$ (or any equation obtained by adding a multiple of $59 \cdot 26$ to both sides of this equation).

13 The first of the given matrices is equivalent to the matrix $\begin{bmatrix} 2 & 3 \\ 0 & 3 \end{bmatrix}$ obtained by adding the second row to the first. The algorithm of Section 7 shows that this new matrix is equivalent to $\begin{bmatrix} 1 & 0 \\ 0 & 6 \end{bmatrix}$. The solution of the second example is similar.

14 All entries of the first matrix are even. It follows easily from the definition of linked matrices that if all entries of a matrix are even then so are all entries of any matrix linked to it. Therefore, all entries of any matrix equivalent to the first matrix are even. Therefore, the second matrix is not equivalent to the first matrix.

15 The given matrix is equivalent to $\begin{bmatrix} -1 & 1 \\ 0 & -1 \end{bmatrix}$, and the algorithm of Section 7 shows that this matrix is equivalent to I_2. If one keeps track of the tilts called for by the algorithm, one finds

$$\begin{bmatrix} 1 & 0 \\ 1 & 1 \end{bmatrix}^2 \begin{bmatrix} -1 & 1 \\ 0 & -1 \end{bmatrix} \begin{bmatrix} 1 & 0 \\ 1 & 1 \end{bmatrix}^2 \begin{bmatrix} 1 & -1 \\ 0 & 1 \end{bmatrix} = I_2$$

from which

$$\begin{bmatrix} 1 & 0 \\ 1 & 1 \end{bmatrix}^2 \begin{bmatrix} 1 & -1 \\ 0 & 1 \end{bmatrix} \begin{bmatrix} -1 & 0 \\ 0 & -1 \end{bmatrix} \begin{bmatrix} 1 & 0 \\ 1 & 1 \end{bmatrix}^2 \begin{bmatrix} 1 & -1 \\ 0 & 1 \end{bmatrix} = I_2$$

follows. This equation is of the form $MAN = B$, and it yields an equation of the form $A = M^{-1}BN^{-1}$, namely,

$$\begin{bmatrix} -1 & 0 \\ 0 & -1 \end{bmatrix} = \begin{bmatrix} 1 & 1 \\ 0 & 1 \end{bmatrix} \begin{bmatrix} 1 & 0 \\ -1 & 1 \end{bmatrix}^2 \begin{bmatrix} 1 & 1 \\ 0 & 1 \end{bmatrix} \begin{bmatrix} 1 & 0 \\ -1 & 1 \end{bmatrix}^2$$

which gives $\begin{bmatrix} -1 & 0 \\ 0 & -1 \end{bmatrix}$ as a product of tilts. There are many other ways of writing this matrix as a product of tilts.

16 Since $\begin{bmatrix} -1 & 0 \\ 0 & -1 \end{bmatrix}$ is unimodular, $\begin{bmatrix} -1 & 0 \\ 0 & 1 \end{bmatrix} \sim \begin{bmatrix} -1 & 0 \\ 0 & 1 \end{bmatrix} \begin{bmatrix} -1 & 0 \\ 0 & -1 \end{bmatrix} \sim \begin{bmatrix} 1 & 0 \\ 0 & -1 \end{bmatrix}$. It follows that if a diagonal matrix whose diagonal entries are all ± 1 has a -1 in a position other than the last, then it is equivalent to the matrix obtained by changing the first -1 on the diagonal to 1 and reversing the sign of the next entry on the diagonal (because the row and column operations that transform $\begin{bmatrix} -1 & 0 \\ 0 & \pm 1 \end{bmatrix}$ into $\begin{bmatrix} 1 & 0 \\ 0 & \mp 1 \end{bmatrix}$ can be performed on a 2×2 diagonal matrix within a larger diagonal matrix without altering the other entries of the matrix; all entries in the affected rows and columns outside the 2×2 matrix that is being changed are zero). This operation either leaves the number of minuses on the diagonal unchanged (if the diagonal entry following the first -1 is 1) or reduces it by 2 (if that diagonal entry is -1). Repetition of this operation shows that such a matrix is equivalent either to I (if -1 occurs an even number of times) or to the matrix obtained by changing the last diagonal entry of I to -1 (if it occurs an odd number of times).

No techniques were introduced in this chapter for proving that matrices are *in*equivalent, so no way of proving that $\begin{bmatrix} -1 & 0 & 0 \\ 0 & -1 & 0 \\ 0 & 0 & -1 \end{bmatrix}$ is not unimodular is yet available. (See Chapter 5.) However, since 1×1 matrices are equivalent only if they are equal, $[-1]$ is not equivalent to $[1]$, that is, $[-1]$ is not unimodular.

17 As was noted in Section 4, every unimodular matrix has an inverse. Multiplication of the equation $MAN = I$ on the left by M^{-1} gives $AN = M^{-1}$. Multiplication of this equation on the right by M gives $ANM = I$. Similarly, $NMA = I$. Thus, NM inverts A.

(a) $A^{-1} = \begin{bmatrix} 4 & -7 \\ -5 & 9 \end{bmatrix} \begin{bmatrix} 1 & 0 \\ -1 & 1 \end{bmatrix} = \begin{bmatrix} 11 & -7 \\ -14 & 9 \end{bmatrix}.$

(b) $A^{-1} = \begin{bmatrix} 1 & -1 & 0 \\ 0 & 1 & -1 \\ 0 & 0 & 1 \end{bmatrix} \begin{bmatrix} 1 & 0 & 0 \\ -2 & 1 & 0 \\ 0 & -1 & 1 \end{bmatrix} = \begin{bmatrix} 3 & -1 & 0 \\ -2 & 2 & -1 \\ 0 & -1 & 1 \end{bmatrix}.$

18 Let T_i be the tilt obtained by adding column i to column $i+1$ of I_5. Then $S_2 = T_1$. Multiplication of any matrix with 5 columns on the right by S_i adds the first column to column i and leaves the other columns unchanged. The operation of multiplication on the right by S_4 therefore carries a row matrix with entries a, b, c, d, e to $a, b, c, a+d, e$. This operation can be achieved by multiplying on the right by T_3^{-1} to get $a, b, c, d-c, e$, then multiplying on the right by S_3 to get $a, b, a+c, d-c, e$, then multiplying on the right by T_3 to get $a, b, a+c, a+d, e$, and then multiplying on the right by S_3^{-1} to get $a, b, c, a+d, e$. In other words, multiplication of any matrix with 5 columns on the right by S_4 is the same as multiplying it on the right by $T_3^{-1}S_3T_3S_3^{-1}$. Therefore $S_4 = T_3^{-1}S_3T_3S_3^{-1}$. In a similar way, $S_{i+1} = T_i^{-1}S_iT_iS_i^{-1}$ for all i, which shows that S_{i+1} is unimodular if S_i is. It fact, it gives explicit formulas for S_i as a product of tilts: $S_2 = T_1$; $S_3 = T_2^{-1}T_1T_2T_1^{-1}$; $S_4 = T_3^{-1}T_2^{-1}T_1T_2T_1^{-1}T_3T_1T_2^{-1}T_1^{-1}T_2$. Finally, S_5 is a product of 22 tilts that can easily be written down. The same method enables one to write any matrix J obtained from I_n by changing one of its 0s to 1 as a product of tilts. Therefore, $J^{\pm1}A$ and $AJ^{\pm1}$ are equivalent to A.

20 Revise step 2 as follows. If one entry is positive and the other is not, set a counter equal to 1. As long as twice the counter times the positive entry plus the other entry is negative, double the counter. Add the counter times the positive entry to the other one. Revise step 3 as follows. If the lesser entry is negative, proceed as before. If it is positive, set a counter equal to 1. As long as the greater entry is greater than twice the counter times the lesser entry, double the counter. Subtract the counter times the lesser entry from the greater one.

Chapter 3

1 $X = \begin{bmatrix} 4 \\ 2 \\ a \end{bmatrix}$, where a is any integer.

2 The given problem is to find all 3×1 matrices X, with entries x, y, z for which $AX = 0$, where $A = \begin{bmatrix} 3 & 1 & 2 \\ 1 & 1 & -1 \end{bmatrix}$. This problem is solved by finding

$$\begin{bmatrix} 1 & 0 \\ 1 & 1 \end{bmatrix}\begin{bmatrix} 3 & 1 & 2 \\ 1 & 1 & -1 \end{bmatrix}\begin{bmatrix} 1 & 1 & -3 \\ -2 & -1 & 5 \\ 0 & -1 & 2 \end{bmatrix} = \begin{bmatrix} 1 & 0 & 0 \\ 0 & 1 & 0 \end{bmatrix}$$

and setting $X = \begin{bmatrix} 1 & 1 & -3 \\ -2 & -1 & 5 \\ 0 & -1 & 2 \end{bmatrix} Z$, where Z is the most general solution of $\begin{bmatrix} 1 & 0 & 0 \\ 0 & 1 & 0 \end{bmatrix} Z = 0$. Thus $x = -3a$, $y = 5a$, $z = 2a$, where a is an arbitrary integer.

3 The algorithm of Chapter 2 gives $\begin{bmatrix} 1 & 0 \\ 2 & 1 \end{bmatrix}\begin{bmatrix} 2 & 1 \\ -1 & 1 \end{bmatrix}\begin{bmatrix} 1 & -1 \\ -1 & 2 \end{bmatrix} = \begin{bmatrix} 1 & 0 \\ 0 & 3 \end{bmatrix}$. Thus, $X = \begin{bmatrix} 1 & -1 \\ -1 & 2 \end{bmatrix} Z$, where Z is any solution of $\begin{bmatrix} 1 & 0 \\ 0 & 3 \end{bmatrix} Z = \begin{bmatrix} 1 & 0 \\ 2 & 1 \end{bmatrix}\begin{bmatrix} 9 & 8 \\ -3 & -1 \end{bmatrix} = \begin{bmatrix} 9 & 8 \\ 15 & 15 \end{bmatrix}$. There is exactly one solution $Z = \begin{bmatrix} 9 & 8 \\ 5 & 5 \end{bmatrix}$ of this equation. Therefore, there is exactly one solution $X = NZ = \begin{bmatrix} 4 & 3 \\ 1 & 2 \end{bmatrix}$ of the given equation.

4 Numbers a, b, c, d in the equation of the desired plane can be described as entries of a 4×1 matrix X for which $AX = 0$, where $A = \begin{bmatrix} 1 & 2 & 0 & 1 \\ 0 & 2 & 1 & 1 \\ 1 & 1 & 1 & 1 \end{bmatrix}$. Now by the method of Chapter 2

$$\begin{bmatrix} 1 & 0 & 0 \\ -1 & 1 & 1 \\ -3 & 2 & 3 \end{bmatrix} A \begin{bmatrix} 1 & -1 & 1 & -1 \\ 0 & 4 & -1 & -1 \\ 0 & 2 & 0 & -1 \\ 0 & -7 & 1 & 3 \end{bmatrix} = \begin{bmatrix} 1 & 0 & 0 & 0 \\ 0 & 1 & 0 & 0 \\ 0 & 0 & 1 & 0 \end{bmatrix}$$

(a rather long calculation to do by hand). The most general solution Z of $DZ = MY$ when $Y = 0$ is obviously the 4×1 matrix whose first 3 entries are zero and whose last entry is an arbitrary integer a. Therefore, the most general solution X of $AX = Y$ is

$$X = \begin{bmatrix} -a \\ -a \\ -a \\ 3a \end{bmatrix}.$$

With $a = 1$, this gives $-x - y - z + 3 = 0$, that is, $x + y + z = 3$, as the equation of the desired plane. The direct method gives three equations $a + 2b + d = 0$, $2b + c + d = 0$, $a + b + c + d = 0$. Subtraction of the last from the first gives $b = c$. Subtraction of the second from the first gives $a = c$. Substitution in the last then gives $d = -3a$, from which the desired solution follows.

5 $\begin{bmatrix} 1 & 0 & 0 \\ 0 & 1 & 0 \\ -1 & -1 & 1 \end{bmatrix}\begin{bmatrix} 2 & 3 \\ 2 & 4 \\ 4 & 7 \end{bmatrix}\begin{bmatrix} 2 & -3 \\ -1 & 2 \end{bmatrix} = \begin{bmatrix} 1 & 0 \\ 0 & 2 \\ 0 & 0 \end{bmatrix}$. By inspection, the equation $DZ = MY$, that is, the equation $\begin{bmatrix} 1 & 0 \\ 0 & 2 \\ 0 & 0 \end{bmatrix} Z = \begin{bmatrix} r \\ s \\ -r - s + t \end{bmatrix}$, has a solution Z if and only if two conditions are satisfied: (a) s must be even, and (b) $-r - s + t$ must be zero.

6 $\begin{bmatrix} 1 & 0 \\ -1 & 1 \end{bmatrix}\begin{bmatrix} 2 & 5 \\ 1 & 2 \end{bmatrix}\begin{bmatrix} 3 & -5 \\ -1 & 2 \end{bmatrix} = \begin{bmatrix} 1 & 0 \\ 0 & -1 \end{bmatrix}$. The only solution Z of $DZ = MI$ is $Z = \begin{bmatrix} 1 & 0 \\ 1 & -1 \end{bmatrix}$. Therefore, the only solution is $X = NZ = \begin{bmatrix} 3 & -5 \\ -1 & 2 \end{bmatrix}\begin{bmatrix} 1 & 0 \\ 1 & -1 \end{bmatrix} = \begin{bmatrix} -2 & 5 \\ 1 & -2 \end{bmatrix}$. (Thus, X is the inverse of A—see Exercise 16.)

7 Clearly $x = 3$, $y = 0$ is a solution. The algorithm of Chapter 2 gives

$$[3 \;\; 2]\begin{bmatrix} 1 & -2 \\ -1 & 3 \end{bmatrix} = [1 \;\; 0],$$

which leads to $MY = 9$, $Z = \begin{bmatrix} 9 \\ a \end{bmatrix}$, $X = \begin{bmatrix} 9 - 2a \\ -9 + 3a \end{bmatrix}$, where a is an arbitary integer. The case $a = 3$ is the solution that was found by inspection.

8 Clearly y and z can be given any value and x determined by $x = -2y - 3z$. The procedure of the text gives $X = NZ$, where $N = \begin{bmatrix} 1 & -1 & 0 \\ 0 & 2 & -3 \\ 0 & -1 & 2 \end{bmatrix}$ and $Z = \begin{bmatrix} 0 \\ s \\ t \end{bmatrix}$, s and t being arbitrary integers. Thus $x = -s$, $y = 2s - 3t$, $z = -s + 2t$. This solution can be shown to agree with the previous one by showing that the matrix $\begin{bmatrix} 2 & -3 \\ -1 & 2 \end{bmatrix}$ is unimodular with inverse $\begin{bmatrix} 2 & 3 \\ 1 & 2 \end{bmatrix}$. Thus, y and z can be given any desired values by setting $s = 2y + 3z$ and $t = y + 2z$. Once the values of y and z are chosen, the value of x must be $-s = -2y - 3z$.

9 (a) $X = \begin{bmatrix} 3 - 4a & -4b \\ -2 + 3a & 3b \end{bmatrix}$, where a and b are arbitrary integers. (b) No solution. (c) The only solution is $X = \begin{bmatrix} 2 & -3 \\ -1 & 2 \end{bmatrix}$.

10 There is a solution if and only if w is a multiple of 3. When w is a multiple of 3, the most general solution is $x = 2w - 19s$, $y = -(11w/3) + 35s$, where s is an arbitrary integer.

11 The most general solution is $X = \begin{bmatrix} 4t \\ -t \\ 5t \end{bmatrix}$, where t is an arbitrary integer.

12 Find unimodular matrices M and N such that $MAN = D$ is diagonal. The solutions X of $XA = Y$ correspond one-to-one to solutions Z of $ZD = YN$ via $X = ZM$, $Z = XM^{-1}$. In particular, there is a solution if and only if the ith column of YN is a multiple of d_i for $i = 1, 2, \ldots,$ k, and, in the case $n > m$, all columns of YN past the mth column are zero.

13 By the preceding exercise, solutions X of $XA = 0$ correspond one-to-one with solutions Z of $ZD = 0$ via $X = ZM$, $Z = XM^{-1}$. Since $X \neq 0$ if and only if $Z \neq 0$, A is a right zero divisor if and only if D is. Multiplication of Z on the right by D multiplies the ith column of Z by the ith diagonal entry of D if $i \leq m$ where m is the number of rows of D, and multiplies it by zero if $i > m$. Therefore, $Z \neq 0$ and $ZD = 0$ can occur only when a diagonal entry of D is zero or D contains more than m rows, which is true if and only if D contains a zero row.

14 If $A' = MAN$ where M and N are unimodular, and if B is a right inverse of A, then $N^{-1}BM^{-1}$ is a right inverse of A' because $A'N^{-1}BM^{-1} = MANN^{-1}BM^{-1} = MABM^{-1} = MM^{-1} = I$. Because equivalence is a symmetric relation, it follows that A has a right inverse if and only if A' does. If D is diagonal, say with diagonal entries $d_1, d_2, \ldots, d_k$, $DB = I$ has a solution only if the ith row of I is a multiple of d_i for $i = 1, 2, \ldots, k$, and, when $n < m$, only if all rows of I past the nth row are zero. Therefore, n must be at least as great as m and each d_i must be ± 1. Conversely, if D is a diagonal $m \times n$ matrix in which $n \geq m$ and all nonzero entries are ± 1, then D is easily seen to have a right inverse. Since no high diagonal matrix has a right inverse, no high matrix has a right inverse by the first part of the exercise.

15 A diagonal matrix with a left inverse must have a nonzero entry (in fact, an entry ± 1) in each column and therefore must not be wide.

16 If $CA = I$ and $AB = I$, then $C = CAB = B$, so any one left inverse is equal to any one right inverse, and the desired conclusions follow. A square diagonal matrix is invertible if and only if its diagonal entries are all ± 1, in which case it is its *own* inverse, that is, its square is the identity matrix. If A is invertible with inverse A^{-1}, then $A' = MAN$ is invertible with inverse $N^{-1}A^{-1}M^{-1}$ whenever M and N are unimodular.

17 Let J_n be the matrix obtained by reversing the sign of the last entry on the diagonal of I_n. Exercise 16 of the last chapter shows that if D is a diagonal matrix that is invertible but not unimodular then $J_n D$ is unimodular. Let A be invertible, and let D be a diagonal matrix equivalent to A. Since D is invertible, either D is unimodular or $J_n D$ is unimodular. In the first case, A is clearly unimodular. It is to be shown that in the second case $J_n A$ is unimodular, or, what is the same, that $J_n A$ is equivalent to $J_n D$. It is easy to see that if T is a tilt then $J_n T J_n$ is also a tilt (multiplication on the left and right by J_n reverses the signs of the entries in the last row and column, leaving the last diagonal entry and all entries not in the last row or column unchanged) and $J_n^2 = I_n$. If $M = T_1 T_2 \cdots T_k$ is unimodular, then $J_n M J_n = J_n T_1 J_n J_n T_2 J_n J_n T_3 \cdots J_n J_n T_k J_n$ is a product of tilts and therefore unimodular. Thus, the equation $J_n A = J_n M J_n J_n DN$ shows $J_n A$ is equivalent to $J_n D$, as desired.

18 If x, y, z are the entries of Y, a necessary and sufficient condition for the first matrix is $x + y = z$. (The last row of M is -1, -1, 1.) In the case of the second matrix, there are two conditions: $x + z = 2y$ and $y =$ even.

Chapter 4

1 For example

$$\begin{vmatrix} 3 & -3 & 6 & 0 & 0 \\ 3 & 0 & 9 & 6 & 2 \\ 3 & 1 & 11 & 12 & 3 \\ 1 & 3 & 4 & -3 & 3 \\ 1 & -1 & 0 & -5 & -2 \end{vmatrix} = \begin{vmatrix} 3 & 0 & 0 & 0 & 0 \\ 3 & 3 & 3 & 6 & 2 \\ 3 & 4 & 5 & 12 & 3 \\ 1 & 4 & 2 & -3 & 3 \\ 1 & 0 & -2 & -5 & -2 \end{vmatrix} = 3\begin{vmatrix} 3 & 3 & 6 & 2 \\ 4 & 5 & 12 & 3 \\ 4 & 2 & -3 & 3 \\ 0 & -2 & -5 & -2 \end{vmatrix} =$$

$$= 3\begin{vmatrix} 3 & 3 & 6 & 2 \\ 1 & 2 & 6 & 1 \\ 1 & -1 & -9 & 1 \\ 0 & -2 & -5 & -2 \end{vmatrix} = 3\begin{vmatrix} 0 & 6 & 33 & -1 \\ 0 & 3 & 15 & 0 \\ 1 & -1 & -9 & 1 \\ 0 & -2 & -5 & -2 \end{vmatrix} = 3\cdot 1\begin{vmatrix} 6 & 33 & -1 \\ 3 & 15 & 0 \\ -2 & -5 & -2 \end{vmatrix} =$$

$$= 3\begin{vmatrix} 6 & 3 & -1 \\ 3 & 0 & 0 \\ -2 & 5 & -2 \end{vmatrix} = -3\cdot 3\begin{vmatrix} 3 & -1 \\ 5 & -2 \end{vmatrix} = -9\begin{vmatrix} 3 & -1 \\ -1 & 0 \end{vmatrix} = (-9)(-1)^{2+1}(-1)(-1) = 9.$$

2 For example, if the 1 in row 4 of column 1 is used to eliminate the entries above and below it, the given determinant is found to be

$$\begin{vmatrix} 0 & 1 & 4 & 4 & -4 \\ 0 & 1 & 1 & 1 & 4 \\ 0 & -3 & -7 & -8 & 8 \\ 1 & 2 & 3 & 4 & -3 \\ 0 & -3 & -2 & -3 & 8 \end{vmatrix} = (-1)^{4+1}\begin{vmatrix} 1 & 4 & 4 & -4 \\ 1 & 1 & 1 & 4 \\ -3 & -7 & -8 & 8 \\ -3 & -2 & -3 & 8 \end{vmatrix} = -\begin{vmatrix} 1 & 4 & 4 & -4 \\ 1 & 1 & 1 & 4 \\ -3 & -7 & -8 & 8 \\ 0 & 5 & 5 & 0 \end{vmatrix} =$$

$$= -\begin{vmatrix} 1 & 4 & 0 & -4 \\ 1 & 1 & 0 & 4 \\ -3 & -7 & -1 & 8 \\ 0 & 5 & 0 & 0 \end{vmatrix} = -(-1)^{4+2}5\begin{vmatrix} 1 & 0 & -4 \\ 1 & 0 & 4 \\ -3 & -1 & 8 \end{vmatrix} = -(-1)^{3+2}5\cdot(-1)\begin{vmatrix} 1 & -4 \\ 1 & 4 \end{vmatrix} =$$

$$= -5\begin{vmatrix} 1 & -4 \\ 0 & 8 \end{vmatrix} = -5\cdot 8 = -40.$$

3

$$|A| = 1\begin{vmatrix} 0 & 3 & 2 \\ 2 & 0 & 3 \\ 1 & 2 & 0 \end{vmatrix} = \begin{vmatrix} 0 & 3 & 2 \\ 0 & -4 & 3 \\ 1 & 2 & 0 \end{vmatrix} = 1\begin{vmatrix} 3 & 2 \\ -4 & 3 \end{vmatrix} = \begin{vmatrix} 1 & 2 \\ -7 & 3 \end{vmatrix} = \begin{vmatrix} 1 & 0 \\ -7 & 17 \end{vmatrix} = 17.$$

$$|B| = \begin{vmatrix} 0 & 3 & 2 & 0 \\ 0 & 0 & -10 & 1 \\ 1 & 0 & 5 & 0 \\ 0 & 2 & 0 & 3 \end{vmatrix} = 1\begin{vmatrix} 3 & 2 & 0 \\ 0 & -10 & 1 \\ 2 & 0 & 3 \end{vmatrix} = 2\begin{vmatrix} 3 & 1 & 0 \\ 0 & -5 & 1 \\ 2 & 0 & 3 \end{vmatrix} = 2\begin{vmatrix} 0 & 1 & 0 \\ 15 & -5 & 1 \\ 2 & 0 & 3 \end{vmatrix} =$$

$$= 2\cdot(-1)\cdot\begin{vmatrix} 15 & 1 \\ 2 & 3 \end{vmatrix} = -2\begin{vmatrix} 15 & 1 \\ -43 & 0 \end{vmatrix} = -2\cdot(-1)^{2+1}(-43) = -86.$$

$$AB = \begin{bmatrix} 1 & 6 & 9 & 0 \\ 6 & 4 & 0 & 9 \\ 0 & 12 & 4 & 9 \\ 4 & 3 & 2 & 2 \end{bmatrix}.$$

$$|AB| = \begin{vmatrix} 1 & 6 & 9 & 0 \\ 6 & 4 & 0 & 9 \\ 0 & 12 & 4 & 9 \\ 4 & 3 & 2 & 2 \end{vmatrix} = \begin{vmatrix} 1 & 6 & 9 & 0 \\ -12 & -5 & -9 & 9 \\ -18 & 3 & -5 & 9 \\ 0 & 1 & 0 & 2 \end{vmatrix} =$$

$$= \begin{vmatrix} 1 & 6 & 9 & -12 \\ -12 & -5 & -9 & 19 \\ -18 & 3 & -5 & 3 \\ 0 & 1 & 0 & 0 \end{vmatrix} = \begin{vmatrix} 1 & 9 & -12 \\ -12 & -9 & 19 \\ -18 & -5 & 3 \end{vmatrix} = \begin{vmatrix} -71 & -15 & -12 \\ 102 & 29 & 19 \\ 0 & 1 & 3 \end{vmatrix} =$$

$$= \begin{vmatrix} -71 & -15 & 33 \\ 102 & 29 & -68 \\ 0 & 1 & 0 \end{vmatrix} = - \begin{vmatrix} -71 & 33 \\ 102 & -68 \end{vmatrix} = 2 \begin{vmatrix} -71 & 33 \\ -51 & 34 \end{vmatrix} = 2 \begin{vmatrix} -71 & 33 \\ 20 & 1 \end{vmatrix} =$$

$$= 2 \begin{vmatrix} -71 - 660 & 33 \\ 0 & 1 \end{vmatrix} = -2 \cdot 731.$$

Note that $|AB| = |A||B|$.

4 (a) is a special case of (b).

(b) An $n \times n$ permutation matrix can be constructed by choosing any one of n columns as the location of the 1 in the first row, then choosing any one of the remaining $n-1$ columns as the location of the 1 in the second row, any one of the remaining $n-2$ columns as the location of the 1 in the third row, and so forth, until the location of the 1 in the last row is determined as the only column without a 1 in it as yet. Thus, the number of $n \times n$ permutation matrices is $n \cdot (n-1) \cdot (n-2) \cdots 3 \cdot 2 \cdot 1 = n!$.

Since a permutation matrix has one 1 and the rest zeros in the first column, its determinant is ± 1 times the determinant of an $(n-1) \times (n-1)$ permutation matrix. Since there is only one 1×1 permutation matrix and its determinant is 1, the determinant of any permutation matrix is therefore ± 1. The permutation matrix denoted P_1 in Section 7, namely, the matrix obtained by interchanging the first two columns of I_n, has determinant -1, as was shown in Section 7. If P is a permutation matrix, then PP_1 is P with the first two columns interchanged, so it is a permutation matrix. Moreover, $(PP_1)P_1 = P$ and $|PP_1| = -|P|$. Thus, multiplication on the right by P_1 gives a one-to-one correspondence between $n \times n$ permutation matrices with determinant 1 and those with determinant -1. Therefore they occur in equal numbers.

5 The determinant of an $n \times n$ matrix A is a sum of $n!$ terms, with one term for each $n \times n$ permutation matrix. The term corresponding to the permutation matrix P is $|P|$ times the product of the n entries of A in locations where the entry of P is 1.

6 Let P_i be as in Section 7. Then, for any matrix A with n rows, P_iA is A with rows i and $i+1$ interchanged. It was shown in Section 7 that $|P_i| = -1$. Thus, $|P_iA| = -|A|$, as was to be shown. Interchange of rows i and $i+s$ of a matrix can be accomplished by $2s-1$ adjacent interchanges—s to move the ith row to position $i+s$, and $s-1$ to move what began as row $i+s$ from position $i+s-1$ up to position i. Since each of these interchanges changes the sign of the determinant, the sign is reversed $2s-1$ times, and the net effect is to reverse the sign. Analogous arguments apply to interchanges of columns. The first and third rows of the given matrix are equal. Interchanging them reverses the sign of the determinant (by what was just shown) and does not change the determinant (because it does not change the matrix). Thus, the determinant is its own negative, so it must be zero.

7 Addition of row i to row k can be achieved by performing $|i-k|-1$ interchanges of adjacent rows to move row k to the position adjacent to row i, then adding row i to it, then performing $|i-k|-1$ more row interchanges to move it back to its original position. The net effect is to reverse the sign of the determinant $2(k-1)$ times, so the determinant is unchanged. The same method applies in the other cases.

Another approach is to note that the determinant of the matrix obtained by adding row i to row k is a sum of two determinants, one of which is the determinant of the original matrix and the other of which is the determinant of the matrix obtained by replacing row k with row i and leaving the other rows unchanged. The second determinant is zero by the previous exercise.

Yet another approach is to observe that if J is the matrix obtained by adding row i to row k of the identity matrix, then the ith diagonal 1 in J is the only nonzero entry in its row, and deleting its row and column gives an identity matrix, so $|J| = 1$. Therefore, $|JA| = |J||A| = |A|$, as was to be shown.

8 If A and B are square matrices and A is linked to B, then, directly from the definition, the transpose of A is linked to the transpose of B. (For example, if B is obtained from A by adding the first row to the second, then the transpose of B is obtained from the transpose of A by adding the first column to the second.) Therefore, if two matrices are equivalent, then their transposes are equivalent. Any A is equivalent to a diagonal matrix D, so its transpose is equivalent to the transpose of D, which is D. Since A and its transpose are equivalent to the same matrix, they are equivalent to each other and therefore have the same determinant.

9 Let A have determinant 1. Let D be a diagonal matrix equivalent to A. By the definition of the determinant, the product of the diagonal entries of D is 1. Therefore, the diagonal entries of D are all ± 1. Since their product is 1, the number of diagonal entries equal to -1 must be even. Therefore, D is unimodular by Exercise 16 of Chapter 2. Thus, A is unimodular.

10 The entry c_j in the jth row of the jth column of A is the only nonzero entry in its row. Deleting its row and column leaves I_{n-1}. Therefore, $|A| = c_j$.

11 Taking the determinant of the given equation gives $\begin{vmatrix} 2 & 1 & 4 \\ 3 & -2 & 5 \\ 1 & 2 & 3 \end{vmatrix} y = \begin{vmatrix} 2 & 3 & 4 \\ 3 & 0 & 5 \\ 1 & 4 & 3 \end{vmatrix}$. The first determinant is easily found to be -4 and the second to have the same value, so $y = 1$. Similarly, $\begin{vmatrix} 2 & 1 & 4 \\ 3 & -2 & 5 \\ 1 & 2 & 3 \end{vmatrix} x = \begin{vmatrix} 3 & 1 & 4 \\ 0 & -2 & 5 \\ 4 & 2 & 3 \end{vmatrix}$ leads to $x = -1$. In the same way, $z = 1$.

12 Let M_i be the matrix obtained by replacing the ith column of I_n with the column $x_1, x_2, \ldots, x_n$ and let A denote the matrix of coefficients of the linear substitution in this problem. Then AM_i is the matrix, call it B_i, obtained by replacing the ith column of A with the column $c_1, c_2, \ldots, c_n$. The determinant of M_i is x_i. Therefore, if $|A| \neq 0$, $x_i = |B_i|/|A|$.

13 Let A be a given $n \times n$ matrix with $|A| \neq 0$ and let C be the jth column of A^{-1}. Since $AA^{-1} = I_n$, AC is equal to the jth column of I_n. By Cramer's rule, the desired ith entry of C is therefore equal to $|B_i|/|A|$ where B_i is the matrix obtained by replacing the ith column of A with the jth column of I_n. The only nonzero entry in the ith column of B_i is a 1 in the jth row, so $|B_i|$ is $(-1)^{i+j}$ times the determinant of the $(n-1) \times (n-1)$ matrix obtained by deleting the ith column and the jth row of B_i. This $(n-1) \times (n-1)$ matrix is the same as the matrix obtained by deleting the ith column and the jth row of A. Thus, the desired entry is $(-1)^{i+j}|A_{ji}|/|A|$, where A_{ji} is the $(n-1) \times (n-1)$ matrix obtained by deleting row j and column i of A.

14 By Exercise 16 of Chapter 2, $-I_n$ is unimodular when n is even. When n is odd, $-I_n$ is *not* unimodular (something that can not be proved using the methods of Chapter 2) because its determinant is not 1.

15 As is seen in Exercise 6 of the Supplementary Unit of Chapter 5, if A is a square matrix with a nonzero determinant, then the number of elements in G_A is the absolute value of $|A|$. Since $A \sim B$ implies that G_A and G_B are isomorphic, the desired conclusion follows.

Supplementary Unit: Exterior Algebra

1 (a) When there are n variables, there are 2^n atoms, one for each subset—including the empty subset—of the set of variables. (b) $1, a, b, c, d, a \wedge b, a \wedge c, a \wedge d, b \wedge c, b \wedge d, c \wedge d, a \wedge b \wedge c, a \wedge b \wedge d, a \wedge c \wedge d, b \wedge c \wedge d, a \wedge b \wedge c \wedge d$. (c) Since a sum of two exterior polynomials in canonical form is easily written in canonical form, it will suffice to prove that every monomial is equal to an exterior polynomial in canonical form. Since an integer times an exterior polynomial in canonical form is in canonical form, it will suffice to show that any product of variables is equal to an exterior polynomial in canonical form. By virtue of $x \wedge y = -y \wedge x$, a product of variables is equal, as an exterior polynomial, to ± 1 times any other product of the same variables in a different order. In particular, it is equal to ± 1 times a product in which the variables appear in order. If some variable occurs twice, the product is equal to 0, which is in canonical form, by virtue of $x \wedge x = 0$. If no variable occurs twice, the product is equal to ± 1 times an atom, which is in canonical form.

2 (a) If two exterior polynomials in canonical form are equal, say $P_1 = P_2$, then adding $(-1)P_2$ to both sides of the equation and combining terms with the same atoms gives an equation of the form $P_3 = 0$ in which P_3 is in canonical form. If P_1 and P_2 are not equal, the integer multipliers in P_3 are not all zero. Let T be the first term in P_3 in which the integer multiplier is not zero, let U be the product of all the variables that do not occur in T, and let both sides of the equation $P_3 = 0$ be multiplied on the right by U. This gives an equation of the form $Q = 0$, in which all terms in Q after the first are equal to zero because they contain a product of variables in which at least one variable occurs at least twice, and in which the first term of Q is a nonzero integer times a product of all the variables in some order. Thus it gives an equation of the desired form, except that the variables may not be in order. Since interchange of two variables changes the sign of the integer multiplier but does not change the fact that the integer multiplier is nonzero, the desired conclusion follows. (b) Part (a) shows that what is to be proved is that the exterior polynomial ax, where a is an integer and x is the one variable, is zero as an exterior polynomial only if $a = 0$. By definition, this means that it is impossible to give a finite sequence of noncommutative polynomials in x in which the first one is ax, the last one is 0, and each one in the sequence can be obtained from its predecessor by applying the rule $x \wedge x = 0$. Since this rule merely allows one to insert or delete terms containing $x \wedge x$, it can have no effect on the term ax, so the only way the last noncommutative polynomial can have no term ax is for the original one to have no such term, i.e., $a = 0$. (c) What is to be shown is that if a sequence of noncommutative polynomials in x and y starts with $ax \wedge y$, ends with 0, and at each step uses one of the rules $x \wedge x = 0$, $y \wedge y = 0$, or $x \wedge y = -y \wedge x$, then $a = 0$. Let $P_0 = ax \wedge y$, P_1, P_2, …, $P_k = 0$ be such a sequence. For each $i = 0, 1, \ldots, k$, let $v(P_i)$ be the coefficient of $x \wedge y$ in P_i minus the coefficient of $y \wedge x$. Then $v(P_0) = a$, $v(P_i) = v(P_{i+1})$, and $v(P_k) = 0$. Therefore, $a = 0$. (d) Use the argument of part (c) with the definition of v changed to $c_{xyz} + c_{yzx} + c_{zxy} - c_{zyx} - c_{yxz} - c_{xzy}$, where c_{xyz} is the coefficient of $x \wedge y \wedge z$, c_{yzx} the coefficient of $y \wedge z \wedge x$, etc. It must be shown that use of any one of the identities $x \wedge y = -y \wedge x$, etc., to change P_i to P_{i+1} leaves $v(P_i) = v(P_{i+1})$. This can be done by a simple enumeration of cases. (e) To generalize the argument of part (d), what needs to be shown is that the $n!$ possible orders of the n variables can be divided into two subsets—xyz, yzx, zxy and zyx, yxz, xzy in the case of three variables—in such a way that the interchange of any two variables carries any order in one set into an order in the other set. This proof can be given in various ways. See, for example, Exercise 5 of Chapter 4. Note that this is, in effect, the formula for the determinant of a square matrix.

3 (a) $A^\wedge = \begin{bmatrix} 1 & 0 & 0 & 0 \\ 0 & 1 & 0 & 0 \\ 0 & 1 & 1 & 0 \\ 0 & 1 & 2 & 0 \\ 0 & 0 & 0 & 1 \\ 0 & 0 & 0 & 2 \\ 0 & 0 & 0 & 1 \\ 0 & 0 & 0 & 0 \end{bmatrix}$. (b) $A^\wedge = \begin{bmatrix} 1 & 0 & 0 & 0 \\ 0 & 3 & 5 & 0 \\ 0 & 2 & 4 & 0 \\ 0 & 0 & 0 & 2 \end{bmatrix}$.

$$(c)\quad A^\wedge = \begin{bmatrix} 1 & 0 & 0 & 0 & 0 & 0 & 0 & 0 \\ 0 & 0 & 3 & 1 & 0 & 0 & 0 & 0 \\ 0 & 2 & 0 & -1 & 0 & 0 & 0 & 0 \\ 0 & 2 & 1 & 0 & 0 & 0 & 0 & 0 \\ 0 & 0 & 0 & 0 & -6 & -2 & -3 & 0 \\ 0 & 0 & 0 & 0 & -6 & -2 & -1 & 0 \\ 0 & 0 & 0 & 0 & 2 & 2 & 1 & 0 \\ 0 & 0 & 0 & 0 & 0 & 0 & 0 & -4 \end{bmatrix}.$$

4 Let (x) represent the original variables of the linear substitution A; let (u) represent the new variables of the linear substitution A, which are the same as the original variables of the linear substitution B; and let (t) represent the new variables of the linear substitution B. Both $(AB)^\wedge$ and $A^\wedge B^\wedge$ describe the linear substitution which expresses atoms of the exterior algebra on the variables (x) in terms of atoms of the exterior algebra on the variables (t). Since they describe the same linear substitution, they have the same matrix of coefficients.

5 (a) $A^{(i)}$ has as many rows as there are atoms containing i variables in the exterior algebra on m variables, which is to say it has $\binom{m}{i}$ rows, where the binomial coefficient $\binom{m}{i}$ is the number $m!/i!(m-i)!$ of i-element subsets of a set of m elements. The number of columns is $\binom{n}{i}$. (b) The matrix $(AB)^{(i)} = A^{(i)}B^{(i)}$ describes the linear substitution which describes the i-atoms of the original variables of A in terms of the i-atoms of the new variables of B. (c) Write A as a product of tilts and one diagonal matrix, say $A = A_1A_2\cdots A_N$. Then $A^{(n)} = A_1^{(n)}A_2^{(n)}\cdots A_N^{(n)}$. Because $x \wedge (x+y) = x \wedge y$, it is easy to see that $A_j^{(n)} = 1$ whenever A_j is a tilt. Therefore, if A is equivalent to a diagonal matrix D, $A^{(n)} = D^{(n)}$. It is clear from the definitions that $D^{(n)} = |D|$; since $|A| = |D|$, the desired conclusion $A^{(n)} = |A|$ follows.

Chapter 5

1

(a) $\begin{bmatrix} 1 & 0 & 0 & 0 \\ 0 & 1 & 0 & 0 \\ 0 & 0 & 2 & 0 \\ 0 & 0 & 0 & -12 \end{bmatrix}$ (b) $\begin{bmatrix} 1 & 0 & 0 \\ 0 & 2 & 0 \\ 0 & 0 & 12 \\ 0 & 0 & 0 \end{bmatrix}$ (c) $\begin{bmatrix} 1 & 0 & 0 & 0 \\ 0 & 1 & 0 & 0 \\ 0 & 0 & 2 & 0 \\ 0 & 0 & 0 & -4 \end{bmatrix}$

2 $A_1 = \begin{bmatrix} 1 & 2 \\ 3 & 4 \\ 5 & 6 \end{bmatrix}$, $\mathcal{M}(A_1, 2) = \{-2, -4\}$; $A_2 = \begin{bmatrix} 1 & 1 \\ 3 & 1 \\ 5 & 1 \end{bmatrix}$, $\mathcal{M}(A_2, 2) = \{-2, -4\}$; $A_3 = \begin{bmatrix} 1 & 0 \\ 3 & -2 \\ 5 & -4 \end{bmatrix}$, $\mathcal{M}(A_3, 2) = \{-2, -4\}$; $A_4 = \begin{bmatrix} 1 & 0 \\ 3 & -2 \\ 2 & -2 \end{bmatrix}$, $\mathcal{M}(A_4, 2) = \{-2\}$; $A_5 = \begin{bmatrix} 1 & 0 \\ 1 & 0 \\ 2 & -2 \end{bmatrix}$, $\mathcal{M}(A_5, 2) = \{0, -2\}$; $A_6 = \begin{bmatrix} 1 & 0 \\ 1 & 0 \\ 1 & -2 \end{bmatrix}$, $\mathcal{M}(A_6, 2) = \{0, -2\}$; $A_7 = \begin{bmatrix} 1 & 0 \\ 1 & 0 \\ 0 & -2 \end{bmatrix}$, $\mathcal{M}(A_7, 2) = \{0, -2\}$; $A_8 = \begin{bmatrix} 1 & 0 \\ 0 & 0 \\ 0 & -2 \end{bmatrix}$, $\mathcal{M}(A_8, 2) = \{0, -2\}$; $A_9 = \begin{bmatrix} 1 & 0 \\ 0 & 2 \\ 0 & -2 \end{bmatrix}$, $\mathcal{M}(A_9, 2) = \{2, -2, 0\}$; $A_{10} = \begin{bmatrix} 1 & 0 \\ 0 & 2 \\ 0 & 0 \end{bmatrix}$, $\mathcal{M}(A_{10}, 2) = \{2, 0\}$.

3 $\mathcal{M}(A, 1) = \{2, 5, 4, 1, -2\}$ with gcd 1. $\mathcal{M}(A, 2) = \{-3, 0, 6, -9, 18\}$ with gcd 3. $\mathcal{M}(A, 3) = \{0\}$. The equivalent strongly diagonal matrix is $\begin{bmatrix} 1 & 0 & 0 \\ 0 & 3 & 0 \\ 0 & 0 & 0 \end{bmatrix}$.

4 The determinant of a unimodular matrix is equal to $\det(I) = 1$. Conversely, let A be a matrix with determinant 1. Let D be the strongly diagonal matrix equivalent to A. Then $\det(D) = 1$. Since $\det(D)$ is an integer multiple of each diagonal entry of D (the entries of D are integers), each diagonal entry of D is ± 1. Since D is strongly diagonal, all diagonal entries are nonnegative, with the possible exception of the last. Since $\det(D) > 0$, the last diagonal entry is nonnegative too. Therefore, all diagonal entries of D are 1, that is, $D = I$. Thus, $A \sim I$, as was to be shown.

5 If A is invertible with inverse B then $\det(A)\det(B) = \det(AB) = \det(I) = 1$. Thus, 1 is an integer multiple of $\det(A)$, which implies $\det(A) = \pm 1$. The only strongly diagonal matrices with determinant ± 1 are I, with determinant 1, and the matrix obtained by changing the last diagonal entry of I to -1, with determinant -1. These two strongly diagonal matrices are invertible and are therefore the only strongly diagonal matrices equivalent to invertible matrices.

6 The last column of CV is zero because it is U^{-1} times the last column of UCV, and this column is zero because it is the last column of a wide diagonal matrix. CW is therefore zero. On the other hand it is the first $i-1$ entries of $N\hat{W}$. Therefore, the first $i-1$ entries of $DN\hat{W}$ are all zero, so they are multiples of d_i. All other entries of $DN\hat{W}$ are multiples of d_j for some $j \geq i$, so they are multiples of d_i. Therefore, all entries of $E\hat{W} = MDN\hat{W}$ are multiples of d_i. Each entry of $\hat{W}e_i$ (a matrix multiplication when e_i is regarded as a 1×1 matrix) is a multiple of the corresponding entry of $E\hat{W}$ because e_i is a multiple of the first i diagonal entries e_j of E, and the remaining entries are all zero by the definition of $\hat{W}$. Thus, all entries of the last column of V times e_i are multiples of d_i, from which it follows that all entries of the last column of $V^{-1}V$ times e_i are multiples of d_i. But e_i is an entry of the last column of $I_m = V^{-1}V$ times e_i.

Supplementary Unit: Finitely Generated Abelian Groups

1 Given any 2×1 matrix $a = \begin{bmatrix} r \\ s \end{bmatrix}$, division can be used to write the entries of a in the form $r = 2q + u$ and $s = 3q' + v$, where $u = 0$ or 1 and $v = 0$, 1, or 2. Then $\begin{bmatrix} r \\ s \end{bmatrix} - \begin{bmatrix} u \\ v \end{bmatrix} = A\begin{bmatrix} q \\ q' \end{bmatrix}$, so every element of G_A is represented by one of the six elements $\begin{bmatrix} u \\ v \end{bmatrix}$. Each of these six elements is a multiple of $\begin{bmatrix} 1 \\ 1 \end{bmatrix}$, namely,

$$\begin{bmatrix} 1 \\ 1 \end{bmatrix} \equiv \begin{bmatrix} 1 \\ 1 \end{bmatrix}, \begin{bmatrix} 2 \\ 2 \end{bmatrix} \equiv \begin{bmatrix} 0 \\ 2 \end{bmatrix}, \begin{bmatrix} 3 \\ 3 \end{bmatrix} \equiv \begin{bmatrix} 1 \\ 0 \end{bmatrix}, \begin{bmatrix} 4 \\ 4 \end{bmatrix} \equiv \begin{bmatrix} 0 \\ 1 \end{bmatrix}, \begin{bmatrix} 5 \\ 5 \end{bmatrix} \equiv \begin{bmatrix} 1 \\ 2 \end{bmatrix}, \begin{bmatrix} 6 \\ 6 \end{bmatrix} \equiv \begin{bmatrix} 0 \\ 0 \end{bmatrix}.$$

These six elements of G_A are distinct because $\begin{bmatrix} k \\ k \end{bmatrix}$ is divisible on the left by A if and only if k is divisible by both 2 and 3, so $\begin{bmatrix} k \\ k \end{bmatrix} \equiv \begin{bmatrix} l \\ l \end{bmatrix}$ mod A if and only if $k - l$ is a multiple of 6.

2 As in the previous exercise, it is easy to see that the elements x and y of G_A represented by the matrices $\begin{bmatrix} 1 \\ 0 \end{bmatrix}$ and $\begin{bmatrix} 0 \\ 1 \end{bmatrix}$, respectively, have the desired properties. (Other choices of x and y are possible.)

3 The algorithm of Chapter 2 gives $MBN = A$, where $M = \begin{bmatrix} 1 & 0 \\ -3 & 1 \end{bmatrix}$ and $N = \begin{bmatrix} 1 & -2 \\ 0 & 1 \end{bmatrix}$. By the solution of the problem of division on the left by B, $a - b = Bc$ for some c if and only if $M(a - b) = Ac'$ for some c'. In other words, $a \equiv b \bmod B$ if and only if $Ma \equiv Mb \bmod A$. Thus, the elements x' and y' of G_A represented by $M^{-1}\begin{bmatrix} 1 \\ 0 \end{bmatrix} = \begin{bmatrix} 1 \\ 3 \end{bmatrix}$ and $M^{-1}\begin{bmatrix} 0 \\ 1 \end{bmatrix} = \begin{bmatrix} 0 \\ 1 \end{bmatrix}$ have the desired properties.

4 Given $\begin{bmatrix} a_1 \\ a_2 \\ a_3 \end{bmatrix}$, let b_2 be the remainder when a_2 is divided by 3, that is, $a_2 = 3q + b_2$ where q is an integer and b_2 is 0, 1, or 2. Then $\begin{bmatrix} a_1 \\ a_2 \\ a_3 \end{bmatrix} = \begin{bmatrix} 0 \\ b_2 \\ a_3 \end{bmatrix} + A \begin{bmatrix} a_1 \\ q \end{bmatrix}$, so any given a is congruent mod A to one of the required form. If two matrices in the required form are equivalent, then their difference has first entry 0 and has second entry greater than -3 and less than 3. Such a matrix is divisible on the left by A only if it is zero.

5 *G_A is finite if and only if each row of the diagonal matrix A contains a nonzero entry.* Proof: If the ith row of A is a row of zeros, then the infinite number of $m \times 1$ matrices which are 0 in all entries but the ith are all distinct mod A because $a - b = Ac$ has no solution for such matrices a and b unless $a = b$. Therefore, G_A is infinite in this case. Conversely, if each row contains a nonzero entry, say the ith row contains the entry $c_i \neq 0$ (necessarily in the ith column), then two $m \times 1$ matrices are congruent mod A if and only if, for each i, their ith entries leave the same remainders when divided by c_i. Since there are just $|c_i|$ possible remainders, there are precisely $|c_1 c_2 \cdots c_m|$ elements in G_A.

6 Find a diagonal matrix D equivalent to A. G_A is finite if and only if D has a nonzero entry in each row. (In other words, G_A is finite if and only if the rank of A is equal to the number of its rows.) When this is the case, the number of elements G_A contains is equal to the absolute value of the product of the diagonal entries of D. This number is the gcd of the $m \times m$ minors of A, where m is the number of rows of A.

7 Suppose $MAN = B$ where M and N are unimodular. A one-to-one correspondence between elements of G_A and G_B is established by letting the element of G_A represented by a correspond to the element of G_B represented by Ma. This rule assigns congruence classes mod B to congruence classes mod A because $a \equiv b$ mod A means $a - b = Ac$ for some c, which in turn means $Ma - Mb = MANN^{-1}c$, that is, $Ma \equiv Mb$ mod B. The assignment is one-to-one because, in the same way, if $Ma \equiv Mb$ mod B then $M^{-1}Ma \equiv M^{-1}Mb$ mod A, that is, $a \equiv b$ mod A. Since addition is defined both in G_A and G_B by adding representatives, the assignment respects the addition operation.

8 The element represented by $\begin{bmatrix} 1 \\ 1 \end{bmatrix}$, call it x, has order 30, because 30 is the least common multiple of 6 and 10. It follows that $2 \cdot x$ has order 15 and, more generally, that $(30/d) \cdot x$ has order d for any positive divisor d of 30. Thus, there are elements of orders 1, 2, 3, 5, 6, 10, 15, and 30. If y is any element of G_A, $60 \cdot y \equiv 0$ mod A. If k is the order of y and if d is the greatest common divisor of k and 60, then $d = 60a + kb$ for integers a and b, so $d \cdot y = a \cdot (60 \cdot y) + b \cdot (k \cdot y) = a \cdot 0 + b \cdot 0 = 0$ and $d \leq k$. By the definition of the order of y, $d \geq k$, so $d = k$, which shows that k is a divisor of 60. Thus, the order of every element is one of the numbers just listed, unless there is an element of order 60. There are 60 elements in the group, so an element y of order 60 would have the property that every element was of the form $k \cdot y$ for exactly one integer k in the range $1 \leq k \leq 60$. If this were the case, only one element, namely $30 \cdot y$, would have order 2. But the elements $\begin{bmatrix} 3 \\ 0 \end{bmatrix}$, $\begin{bmatrix} 0 \\ 5 \end{bmatrix}$, and $\begin{bmatrix} 3 \\ 5 \end{bmatrix}$ represent distinct elements of G_A of order 2, so there are no elements of order 60. This list of 3 includes all elements of order 2.

9 For every y in this group, $12 \cdot y = 0$. Therefore, as in the last exercise, the order of every element divides 12, which shows that every element has order 1, 2, 3, 4, 6, or 12. There are 7 elements of order 2, namely, those represented by matrices of the form $\begin{bmatrix} 3 \\ 0 \\ 0 \end{bmatrix} i + \begin{bmatrix} 0 \\ 3 \\ 0 \end{bmatrix} j + \begin{bmatrix} 0 \\ 0 \\ 2 \end{bmatrix} k$, where i, j, and k are 0 or 1 and they are not all three 0. There are 8 elements of order 3, namely, those represented by matrices of the form $\begin{bmatrix} 2 \\ 0 \\ 0 \end{bmatrix} i + \begin{bmatrix} 0 \\ 2 \\ 0 \end{bmatrix} j$, where i and j are 0, 1, or 2 and at least one is nonzero. The matrix $\begin{bmatrix} 0 \\ 1 \\ 1 \end{bmatrix}$ represents an element of G_A of order 12 because 12 is the least common multiple of 6 and 4.

10 G_A has three elements of order 2, whereas G_B has only one. Therefore, there is no one-to-one correspondence between the two groups which respects addition, because in such a correspondence an element of order k must correspond to an element of order k.

11 Let q be the number of rows in the $m \times n$ strongly diagonal matrix A with diagonal entry ± 1, and let s be the number of zero rows. Then $r = m - q - s$. Take as $c_1, c_2, \ldots, c_r$ the absolute values of the diagonal entries, in order, of A that are greater than 1 in absolute value. Take as the elements $y_1, y_2, \ldots, y_s$ in the description of G_A the elements represented by the columns of the $m \times s$ matrix that is 0 in its first $m - s$ rows and I_s in its last s rows. Finally, let $x_1, x_2, \ldots, x_r$ be the elements represented by the columns of the $m \times r$ matrix that is 0 in its first q rows and its last s rows and is I_r in the intervening r rows.

12 Let A be a strongly diagonal matrix, and let s be the number of zero rows of A. It will be shown first that s is determined by the group G_A. In fact, s is the largest integer t such that G_A contains t elements $y_1, y_2, \ldots, y_t$ with the property that $j_1 y_1 + j_2 y_2 + \cdots + j_t y_t = 0$ only when the js are all zero. Clearly, G_A contains s such elements, namely, the elements represented by the columns of the $m \times s$ matrix P whose first $m - s$ rows are zero and whose last s rows are I_s. (An $m \times 1$ matrix whose first $m - s$ entries are all zero is $\equiv 0 \bmod A$ if and only if its last s entries are all zero.) Suppose now that G_A contains t such elements, and let P be an $m \times t$ matrix whose t columns represent $y_1, y_2, \ldots, y_t$, respectively. Let P_1 be the first $m - s$ rows of P, and let P_2 be its last s rows. If P_2 were a left zero divisor, there would be a nonzero $t \times 1$ matrix j such that $P_2 j = 0$. In that case, when d is a positive integer that is a multiple of all the nonzero elements of A, it would follow that $Pjd \equiv 0 \bmod A$, since all entries of $P_1 jd$ are divisible by all the nonzero entries of A, contrary to the choice of P (because the $t \times 1$ matrix jd is nonzero). Therefore P_2 must not be a left zero divisor, so P_2 must not be wide; that is, t can be at most s, as was to be shown.

Let c be an integer greater than 1, and let u be the number of nonzero diagonal entries of A that are divisible by c. It will be shown that u is determined by G_A as the maximum number t of elements $x_1, x_2, \ldots, x_t$ of G_A with the properties (1) $cx_j = 0$ for $j = 1, 2, \ldots, t$ and (2) $j_1 x_1 + j_2 x_2 + \cdots + j_t x_t = 0$ only when the js are all divisible by c. Let P be the $m \times u$ matrix which is zero in its last s rows, zero in its first $m - s - u$ rows, and in the remaining u rows is the $u \times u$ diagonal matrix with entries d_i/c, where, for $i = 1, 2, \ldots, u$, the d_i are the nonzero diagonal entries, in order, of A that are divisible by c. The columns of P represent u elements of G_A with properties (1) and (2), so $t \geq u$. On the other hand, $t \leq u$, as can be seen as follows. Assume G_A contains t elements with properties (1) and (2), and let P be an $m \times t$ matrix whose columns represent the elements $x_1, x_2, \ldots, x_t$ of G_A. Let P_1 denote the first $m - u - s$ rows of P, P_2 the middle u rows of P, and P_3 the last s rows of P. Then $P_3 = 0$, because otherwise a column of P times c would have a nonzero entry in one of the last s rows and would therefore not be $\equiv 0 \bmod A$, contary to property (1). If P_2 were a left zero divisor, there would be a nonzero $t \times 1$ matrix j such that $P_2 j = 0$. In fact, since any common factor of the entries of j could be removed, there would be a nonzero matrix j such that $P_2 j = 0$ and such that the entries of j had no common divisor greater than 1. If P_1 is nonexistent—that is, if $m = u + s$—then $Pj = 0$, so $Pj \equiv 0 \bmod A$, contrary to property (2). Otherwise, let b be the last nonzero diagonal entry of A that is not divisible by c. Since every entry of $P_1 jb$ is divisible by b and every entry of $P_2 jb$ and $P_3 jb$ is zero, $Pjb \equiv 0 \bmod A$. Since the entries of jb are not all divisible by c (c does not divide b, and the entries of j have no common divisor), this contradicts condition (2). Therefore, P_2 cannot be a left zero divisor, which implies $t \leq u$, as claimed.

What is to be shown, then, is that if two lists $a_1, a_2, \ldots, a_k$ and $b_1, b_2, \ldots, b_l$ of integers greater than 1 have the properties that (1) each entry other than the first in either list is divisible by its predecessor and (2) the number of entries in one list that are divisible by any $c > 1$ is the same as the number of entries in the other list divisible by that c, then the lists are identical. This follows from the observation that, for $i = 0, 1, \ldots, k-1$, a_{k-i} divides the last $i+1$ entries of the a list, so it must divide at least $i+1$ entries of the b list, so it must divide the *last* $i+1$ entries of the b list, so it must divide b_{l-i}. In the same way, b_{l-i} must divide a_{k-i} for $i = 0, 1, \ldots, l-1$, which shows that the lists coincide.

13 Let D and E be strongly diagonal matrices equivalent to A and B, respectively. Since G_A is isomorphic to G_D and G_B is isomorphic to G_E, G_A is isomorphic to G_B if and only if G_D is isomorphic to G_E. The preceding exercise therefore implies that G_A is isomorphic to G_B if and only if D and E (which can be found by the algorithm of this chapter) have two properties: (1) They have the same number s of zero diagonal entries. (2) The number r of nonzero diagonal entries with absolute value greater than 1 are the same in the two cases, and the two lists of r absolute values of diagonal entries greater than 1 are the same.

14 The first of these two matrices is strongly diagonal and has determinant kl. The second is equivalent to the strongly diagonal matrix whose determinant is kl and whose first diagonal entry is the gcd of k and l. The desired conclusion follows.

15 The preceding exercises show that one can assume without loss of generality that A is strongly diagonal with all diagonal entries positive. Moreover, diagonal entries equal to 1 can be inserted or deleted, and the diagonal entries can be rearranged without changing G_A. A positive integer which is not a power of a prime can be written as a product kl of two relatively prime numbers greater than 1, and a diagonal matrix in which consecutive diagonal entries are 1 and kl is equivalent to the diagonal matrix that is the same except that these two diagonal entries are replaced by k, l. The desired conclusion follows from these observations.

16 Since zero columns can be added or deleted without changing the group, one can assume without loss of generality that A and B are both square. Since diagonal 1s can be inserted or deleted without changing the group, one can assume without loss of generality that A and B have the same size. The groups are isomorphic if and only if the equivalent strongly diagonal matrices are identical. What is to be shown, then, is that if A and B are square diagonal matrices in which the diagonal entries are 0, 1, or a power of a prime, and if they are equivalent, then the diagonal entries are the same except for the order.

Since the number of nonzero diagonal entries is equal to the rank, the number of zero entries is the same in the two matrices. As for the nonzero entries, repeated application of Exercise 14 shows that the diagonal entries of A are characterized by the following property: Let D be the strongly diagonal matrix equivalent to A. Let each diagonal entry of D that is greater than 1 be written as a product of prime powers. The diagonal entries of A greater than 1 coincide with the prime powers obtained in this way. Since the same computation gives the diagonal entries of B, the desired conclusion follows.

17 To say that a group G is finitely generated means that there is a finite set $x_1, x_2, \ldots, x_n$ of elements of G such that every element of G can be expressed in the form $k_1 \cdot x_1 + k_2 \cdot x_2 + \cdots + k_n \cdot x_n$ for some integers $k_1, k_2, \ldots, k_n$. The rigorous proof of the desired proposition involves some philosophical questions about what constitutes an acceptable description of a group. Certainly one should be able, given xs and ks, to determine whether $k_1 \cdot x_1 + k_2 \cdot x_2 + \cdots + k_n \cdot x_n = 0$. A natural requirement is that there be given a finite number of such relations which imply them all. In this case, if m is the number of given relations and A is the matrix whose rows are the ks of the given relations, then the given group is isomorphic to G_A.

Chapter 6

1 (a) $\begin{bmatrix} \frac{1}{2} & 0 & 0 \\ 0 & 1 & 0 \\ 0 & 0 & 1 \end{bmatrix}$. (b) $\begin{bmatrix} \frac{1}{3} & 0 & 0 \\ 0 & 1 & 0 \\ 0 & 0 & 3 \end{bmatrix}$.

2 $MAN = D$ where $M = \begin{bmatrix} 1 & 0 \\ -4 & 1 \end{bmatrix}$, $N = \begin{bmatrix} 1 & -1 & -1 \\ 0 & 5 & 2 \\ 0 & -3 & -1 \end{bmatrix}$, and $D = \begin{bmatrix} 1 & 0 & 0 \\ 0 & 3 & 0 \end{bmatrix}$. The most general solution of $DZ = MI_2$, that is, of $\begin{bmatrix} 1 & 0 & 0 \\ 0 & 3 & 0 \end{bmatrix} Z = \begin{bmatrix} 1 & 0 \\ -4 & 1 \end{bmatrix}$, is $Z = \begin{bmatrix} 1 & 0 \\ -\frac{4}{3} & \frac{1}{3} \\ a & b \end{bmatrix}$, where a and b are arbitrary rational numbers. Therefore, the most general right inverse is $X = NZ =$

$$\begin{bmatrix} 1 & -1 & -1 \\ 0 & 5 & 2 \\ 0 & -3 & -1 \end{bmatrix} \begin{bmatrix} 1 & 0 \\ -\frac{4}{3} & \frac{1}{3} \\ a & b \end{bmatrix} = \begin{bmatrix} \frac{7}{3} - a & -\frac{1}{3} - b \\ -\frac{20}{3} + 2a & \frac{5}{3} + 2b \\ 4 - a & -1 - b \end{bmatrix}.$$

3 The most general right inverse with rational entries is

$$X = \begin{bmatrix} 1 & -1 & -1 \\ -1 & 2 & 1 \\ 2 & -4 & -1 \end{bmatrix} \begin{bmatrix} 1 & 0 \\ -2 & 1 \\ a & b \end{bmatrix} = \begin{bmatrix} 3 - a & -1 - b \\ -5 + a & 2 + b \\ 10 - a & -4 - b \end{bmatrix}$$

where a and b are arbitrary rational numbers. The same computation gives the most general right inverse with integer entries, but in this case a and b are integers. This matrix is wide, so it has *no* left inverse.

4 Multiplication of a matrix with integer entries on either the left or the right by a tilt gives a matrix with integer entries. Therefore, a matrix linked to a matrix with integer entries has integer entries, and the desired conclusion follows.

5 Suppose A has a left inverse, say $CA = I$. If $AX = 0$, then $CAX = C0$, that is, $X = 0$. Therefore, A is not a left zero divisor.

6 $B = MA$ where M is I with one entry off the diagonal changed from 0 to a nonzero rational number. The 1 on the diagonal in the same row as the newly nonzero entry of M is the only nonzero entry in its column, and reduction of the determinant using this observation shows that its determinant is equal to the determinant of an identity matrix. Therefore $|M| = 1$ and $|B| = |M||A| = |A|$.

7 If two matrices of rational numbers are equivalent, then clearly they are equivalent in the wider sense. Therefore, any matrix with rational entries is equivalent in the wider sense to a strongly diagonal matrix. Moreover, if $a \neq 0$ then the diagonal matrix $\begin{bmatrix} a & 0 \\ 0 & b \end{bmatrix}$ is equivalent in the wider sense to the matrices $\begin{bmatrix} a & 1 \\ 0 & b \end{bmatrix}$, $\begin{bmatrix} 0 & 1 \\ -ab & b \end{bmatrix}$, $\begin{bmatrix} 1 & 1 \\ b - ab & b \end{bmatrix}$, $\begin{bmatrix} 1 & 0 \\ b - ab & ab \end{bmatrix}$, $\begin{bmatrix} 1 & 0 \\ 0 & ab \end{bmatrix}$. It follows that a strongly diagonal matrix of rank r is equivalent in the wider sense to a matrix whose first $r - 1$ diagonal entries are all 1, whose rth diagonal entry is nonzero, and whose remaining entries are all zero. If $r < n$, then, because $\begin{bmatrix} a & 0 \end{bmatrix}$ is equivalent in the wider sense to $\begin{bmatrix} 1 & 0 \end{bmatrix}$ when $a \neq 0$, this matrix is equivalent in the wider sense to the matrix with r diagonal 1s and the rest 0s. The same is true when $r < m$ because $\begin{bmatrix} a \\ 0 \end{bmatrix}$ is equivalent in the wider sense to $\begin{bmatrix} 1 \\ 0 \end{bmatrix}$ when $a \neq 0$.

Adding a rational multiple of a row or column to an adjacent row or column does not change the determinant of a matrix, so square matrices that are equivalent in the wider sense have equal determinants. Since, as was just shown, an $n \times n$ matrix A of rank n is equivalent in the wider sense to a diagonal matrix in which all diagonal entries but the last are 1, such a matrix A is equivalent in the wider sense to the matrix whose first $n-1$ diagonal entries are 1 and whose last diagonal entry is $|A|$. It follows that two $n \times n$ matrices of rank n are equivalent in the wider sense if and only if they have the same determinant.

8 By the associativity of matrix multiplication, $ABB^{-1}A^{-1} = AI_nA^{-1} = AA^{-1} = I_n$. Therefore, $B^{-1}A^{-1}$ is a right inverse of AB. Since AB is square, it must therefore be invertible, and $B^{-1}A^{-1}$ must be its inverse.

9 Since $|A|$ is equal to the product of the diagonal entries of D, $|A| \neq 0$ implies that the diagonal entries of D are all nonzero, so D is invertible and has the inverse described in the statement of the exercise. Then, by the preceding exercise, $M^{-1}DN^{-1}$ is invertible and has inverse $ND^{-1}M$, as was to be shown.

10 By the algorithm of Chapter 2,

$$\begin{bmatrix} 1 & 0 & 0 \\ 0 & 1 & -2 \\ -1 & -1 & 3 \end{bmatrix} \begin{bmatrix} 1 & 4 & 1 \\ 2 & 10 & 5 \\ 1 & 4 & 2 \end{bmatrix} \begin{bmatrix} 1 & -1 & 2 \\ 0 & 0 & -1 \\ 0 & 1 & 2 \end{bmatrix} = \begin{bmatrix} 1 & 0 & 0 \\ 0 & 1 & 0 \\ 0 & 0 & 2 \end{bmatrix}.$$

Therefore, $A^{-1} = ND^{-1}M = \begin{bmatrix} 0 & -2 & 5 \\ \frac{1}{2} & \frac{1}{2} & -\frac{3}{2} \\ -1 & 0 & 1 \end{bmatrix}$.

11 D has a left inverse if and only if $m \geq n$ and each of the n diagonal entries of D is nonzero. In this case, a left inverse of D is the $n \times m$ diagonal matrix whose diagonal entries are the reciprocals of the diagonal entries of D, in order. D has a right inverse if and only if $m \leq n$ and each of the m diagonal entries of D is nonzero. In this case, a right inverse of D is given by the same description—the $n \times m$ diagonal matrix whose diagonal entries are the reciprocals of the diagonal entries of D.

12 If D has a left inverse, say $ED = I$, then $NEMA = NEMANN^{-1} = NEDN^{-1} = NN^{-1} = I$, so NEM is a left inverse of A. Similarly, if D has a right inverse E, NEM is a right inverse of A.

13 If the entries of A^{-1} are all integers then $|A^{-1}|$ is an integer, so, because $|A||A^{-1}| = |I_n| = 1$ and because $1 \cdot 1$ and $(-1) \cdot (-1)$ are the only ways to write 1 as a product of two integers, $|A| = |A^{-1}| = \pm 1$. Thus, a necessary condition for A^{-1} to have integer entries is $|A| = \pm 1$. This condition is also sufficient, because if $|A| = \pm 1$ and if $MAN = D$ is diagonal, where M and N are unimodular, then $|D| = \pm 1$, which implies that the diagonal entries of D are all ± 1; therefore, $D^{-1} = D$ has integer entries, so $A^{-1} = ND^{-1}M$ has integer entries. Thus "$|A| = \pm 1$?" tests whether A^{-1} has integer entries.

14 The wide matrix $\begin{bmatrix} 0 & 0 \end{bmatrix}$ has no right inverse. A wide matrix A is always a left zero divisor, that is, there is necessarily a nonzero matrix B such that $AB = 0$. If C is any right inverse of A, then $B + C$ is another right inverse that is different from C.

15 No changes are required. The entry in the ith row and the jth column of A^{-1} is $(-1)^{i+j}|A_{ji}|/|A|$.

Supplementary Unit: Sets and Vector Spaces

1 Saying $A(\mathbf{u}-\mathbf{v}) = 0$ only when $\mathbf{u}-\mathbf{v} = 0$ is the same as saying that A is not a left zero divisor. Thus, the function is one-to-one if and only if A is not a left zero divisor. This is true if and only if the rank of A is n.

2 To say the function is onto is to say that the division problem $A\mathbf{u} = \mathbf{v}$ has at least one solution for each $\mathbf{v} \in \mathbf{Q}^m$. This is true if and only if $D\mathbf{u} = M\mathbf{v}$ has a solution for each $\mathbf{v} \in \mathbf{Q}^m$, where M and N are unimodular matrices such that $MAN = D$ is diagonal. Because rational solutions are allowed, $D\mathbf{u} = M\mathbf{v}$ has a solution if and only if D is zero in all rows in which $M\mathbf{v}$ is zero. Since no row of M is zero ($|M| \neq 0$), this is true for all $\mathbf{v}$ if and only if no row of D is zero. Thus, the function is onto if and only if the rank of A is m.

3 If A is $m \times n$ and B is $p \times q$, then multiplication on the left by A gives a function $\mathbf{Q}^n \to \mathbf{Q}^m$, call it f, and multiplication on the left by B gives a function $\mathbf{Q}^q \to \mathbf{Q}^p$, call it g. The composition $g \circ f$ is defined if and only if the set $\mathbf{Q}^m$ where f has its values coincides with the set $\mathbf{Q}^q$ for which g is defined. When this is the case, the composed function multiplies an element $\mathbf{u}$ of $\mathbf{Q}^n$ on the left by A and multiplies the result on the left by B. By the associativity of matrix multiplication, this is the same as multiplying $\mathbf{u}$ on the left by BA. In short, the composed function is defined when $m = q$—which is when the matrix product BA is defined—and in this case the composed function $g \circ f$ corresponds to BA in the same way that g corresponds to B and f corresponds to A.

4 If f has a left inverse g, and if $f(s_1) = f(s_2)$, then $s_1 = g\left(f(s_1)\right) = g\left(f(s_2)\right) = s_2$, so f is one-to-one. Conversely, suppose f is one-to-one. For each element t of T, there is either exactly one solution s of $f(s) = t$ or there is none. Let s_0 be an arbitrary element of S (S is nonempty), and let $g : T \to S$ be the function that assigns the solution s of $f(s) = t$ to t when there is a solution and otherwise assigns s_0 to t. Then $g\left(f(s)\right) = s$ for all $s \in S$, so g is a left inverse of f.

5 If f has a right inverse g, then for every $t \in T$ the element $g(t) \in S$ has the property that $f\left(g(t)\right) = t$, which shows that f is onto. The converse statement goes by the impressive name Axiom of Choice, and is, as this name implies, so "obvious" it can be taken as an axiom. The idea is that for each element $t \in T$, the set of all elements $s \in S$ for which $f(s) = t$ is nonempty (f is onto); the needed function g chooses, for each t, an element s of the corresponding nonempty set. Then $f(g(t)) = t$ by the definition of g.

6 Let $\mathbf{e}_1, \mathbf{e}_2, \ldots, \mathbf{e}_n$ be the columns, in order, of I_n. The element $\mathbf{u} \in \mathbf{Q}^n$ whose ith entry is u_i for $i = 1, 2, \ldots, n$ is then equal to $\mathbf{u} = \mathbf{e}_1u_1 + \mathbf{e}_2u_2 + \cdots + \mathbf{e}_nu_n$. If $f : \mathbf{Q}^n \to \mathbf{Q}^m$ is a linear function, it follows that $f(\mathbf{u}) = f(\mathbf{e}_1)u_1 + f(\mathbf{e}_2)u_2 + \cdots + f(\mathbf{e}_n)u_n = A\mathbf{u}$, where A is the matrix whose ith column is $f(\mathbf{e}_i)$. The converse, that for any $m \times n$ matrix A the function $\mathbf{Q}^n \to \mathbf{Q}^m$ which carries $\mathbf{u} \to A\mathbf{u}$ is a linear function, follows immediately from the rules of matrix arithmetic.

7 The existence of such a linearly independent set implies the existence of a one-to-one function $\mathbf{Q}^k \to \mathbf{Q}^n$, or, what is the same, of an $n \times k$ matrix of rank k. Since the rank of a matrix is at most the number of rows, $k \leq n$ follows.

8 The existence of an onto linear function $\mathbf{Q}^k \to \mathbf{Q}^n$ implies the existence of an $n \times k$ matrix of rank n, which implies $n \leq k$.

9 The n columns of the identity matrix I_n are a basis of $\mathbf{Q}^n$.

10 If $\mathbf{V}$ is n-dimensional, then there is a one-to-one, onto linear function $\mathbf{Q}^n \to \mathbf{V}$. The (both left and right) inverse function $\mathbf{V} \to \mathbf{Q}^n$ is also linear, as follows immediately from the definition of linearity. If at the same time $\mathbf{V}$ is m-dimensional, there is a one-to-one, onto linear function $\mathbf{Q}^m \to \mathbf{V}$. The composition of this function with the inverse of the previous one gives a one-to-one and onto linear function $\mathbf{Q}^m \to \mathbf{Q}^n$, which implies the existence of an $n \times m$ matrix whose rank is both n and m; therefore, it implies $n = m$.

11 Let n and m be the dimensions of $\mathbf{V}$ and $\mathbf{W}$ respectively. There are then one-to-one, onto linear functions $\mathbf{Q}^n \to \mathbf{V}$ and $\mathbf{Q}^m \to \mathbf{W}$ with linear inverse functions. An onto linear function $\mathbf{V} \to \mathbf{W}$ implies an onto linear function $\mathbf{Q}^n \to \mathbf{V} \to \mathbf{W} \to \mathbf{Q}^m$, which implies the existence of an $m \times n$ matrix of rank m, which implies $n \geq m$. A one-to-one linear function $\mathbf{V} \to \mathbf{W}$ implies a one-to-one linear function $\mathbf{Q}^n \to \mathbf{Q}^m$, which implies $n \leq m$.

Chapter 7

1

(a) $\begin{bmatrix} \frac{1}{6} \\ \frac{1}{6} \\ \frac{1}{6} \end{bmatrix}$ (b) $\begin{bmatrix} \frac{1}{2} & 0 & 0 \\ 0 & \frac{1}{2} & 0 \\ 0 & 0 & \frac{1}{2} \end{bmatrix}$ (c) $\begin{bmatrix} \frac{1}{2} & 0 \\ 0 & \frac{1}{2} \\ 0 & 0 \end{bmatrix}$ (d) $\begin{bmatrix} \frac{1}{2} & 0 \\ 0 & 0 \end{bmatrix}$.

2 The problem is to find an approximate solution of the equations

$$m \cdot 0 + b = 0$$
$$m \cdot 1 + b = 0$$
$$m \cdot 2 + b = 2,$$

that is, an approximate solution $X = \begin{bmatrix} m \\ b \end{bmatrix}$ of $AX = Y$, where $A = \begin{bmatrix} 0 & 1 \\ 1 & 1 \\ 2 & 1 \end{bmatrix}$ and $Y = \begin{bmatrix} 0 \\ 0 \\ 2 \end{bmatrix}$. The mate of A is $\begin{bmatrix} 5 & 3 \\ 3 & 3 \end{bmatrix}^{-1} \begin{bmatrix} 0 & 1 & 2 \\ 1 & 1 & 1 \end{bmatrix} = \begin{bmatrix} -\frac{1}{2} & 0 & \frac{1}{2} \\ \frac{5}{6} & \frac{1}{3} & -\frac{1}{6} \end{bmatrix}$. The best solution is therefore $\begin{bmatrix} 1 \\ -\frac{1}{3} \end{bmatrix}$. Therefore, the desired line is $y = x - \frac{1}{3}$.

3 The problem is to find the best solution $X = \begin{bmatrix} x \\ y \end{bmatrix}$ of $AX = Y$, where $A = \begin{bmatrix} 3 & 2 \end{bmatrix}$ and $Y = 6$. Since the mate of A is $B = \begin{bmatrix} \frac{3}{13} \\ \frac{2}{13} \end{bmatrix}$, the best solution is $X = BY = \begin{bmatrix} \frac{18}{13} \\ \frac{12}{13} \end{bmatrix}$. The solutions (x, y) of $3x + 2y = 6$ form a line in the xy-plane. The method picks out the point on this line for which $x^2 + y^2$ is a minimum, that is, it picks out the point $(\frac{18}{13}, \frac{12}{13})$ on this line that is nearest the origin (the foot of the perpendicular from $(0, 0)$ to the line).

4 The given equation is $AX = Y$ where $A = \begin{bmatrix} 1 & 3 \\ 2 & -1 \\ 1 & -1 \end{bmatrix}$, $X = \begin{bmatrix} a \\ b \end{bmatrix}$, $Y = \begin{bmatrix} 1 \\ 3 \\ 5 \end{bmatrix}$. The desired solution is therefore $X = BY$, where $B = (A^TA)^{-1}A^T = \begin{bmatrix} 6 & 0 \\ 0 & 11 \end{bmatrix}^{-1} \begin{bmatrix} 1 & 2 & 1 \\ 3 & -1 & -1 \end{bmatrix}$, that is, $X = \begin{bmatrix} 2 \\ -\frac{5}{11} \end{bmatrix}$.

5 The problem is to find an approximate solution X of $AX = Y$, where X is the column matrix with entries a, b, c, where

$$A = \begin{bmatrix} 100 & -10 & 1 \\ 0 & 0 & 1 \\ 100 & 10 & 1 \\ 400 & 20 & 1 \end{bmatrix}, \text{ and where } Y = \begin{bmatrix} 110 \\ 0 \\ 95 \\ 400 \end{bmatrix}.$$

The inverse of

$$A^TA = \begin{bmatrix} 180000 & 8000 & 600 \\ 8000 & 600 & 20 \\ 600 & 20 & 4 \end{bmatrix}$$

can be found using the formula for the inverse of a 3×3 matrix (see Exercise 13 of Chapter 4). The result is

$$(A^TA)^{-1} = \begin{bmatrix} \frac{1}{40000} & -\frac{1}{4000} & -\frac{1}{400} \\ -\frac{1}{4000} & \frac{9}{2000} & \frac{3}{200} \\ -\frac{1}{400} & \frac{3}{200} & \frac{11}{20} \end{bmatrix}.$$

Since A^TY is the column matrix with entries 180500, 7850, and 605, the desired solution is

$$\begin{bmatrix} a \\ b \\ c \end{bmatrix} = (A^TA)^{-1}A^TY = \begin{bmatrix} 1.0375 \\ -0.725 \\ -0.75 \end{bmatrix}.$$

6 $t = \frac{10}{7}$ and the minimum value of $\|At - Y\|^2$ is $\frac{3}{7}$. The points At as t varies are the points of a line through the origin and (3, 2, 1). The point on this line nearest to (4, 3, 2) is $(\frac{30}{7}, \frac{20}{7}, \frac{10}{7})$. The perpendicular distance from point to line is $\sqrt{\frac{3}{7}}$.

7 As is noted in Section 4, both P_1 and P_2 satisfy $P_i^2 = P_i$ and $P_i^T = P_i$. Therefore, as follows immediately from the definition, P_i is its own mate. As is seen in the preceding exercise for a particular Y, for any given Y the matrix ABY describes the point on the line through the origin and A that is nearest to Y. Thus, $P_1 = AB$ represents *orthogonal projection* of points Y onto this line. If Y is already on the line, this projection leaves Y unchanged. Therefore $P_1P_1Y = P_1Y$ for all Y, which is the meaning of $P_1^2 = P_1$. For any Y, $-P_2Y = P_1Y - Y$ is the displacement that carries Y to the point on the line through the origin and A nearest to Y. This displacement is perpendicular to the displacement that carries the point on the line nearest to Y to the origin. Thus, the displacement from Y to the origin is decomposed into two displacements, one along the given line and one perpendicular to it. In other words, P_2 is the orthogonal projection of a given displacement onto the plane through the origin perpendicular to the given line. The equation $P_2^2 = P_2$ is therefore the algebraic version of the geometrical statement that this projection does not move points that are already on the plane.

8 By inspection, $\begin{bmatrix} 0 & 1 \\ 0 & 0 \end{bmatrix} = \begin{bmatrix} 1 \\ 0 \end{bmatrix}\begin{bmatrix} 0 & 1 \end{bmatrix}$. Each of the factors of this product is easily seen to have its transpose as its mate. Therefore, the mate of the given matrix is $\begin{bmatrix} 0 \\ 1 \end{bmatrix}\begin{bmatrix} 1 & 0 \end{bmatrix} = \begin{bmatrix} 0 & 0 \\ 1 & 0 \end{bmatrix}$.

9 $MAN = \begin{bmatrix} 1 & 0 & 0 \\ 0 & 1 & 0 \\ 0 & 0 & 0 \end{bmatrix}$ where $M = \begin{bmatrix} 1 & 0 & 0 \\ -2 & 1 & 0 \\ -3 & 1 & 1 \end{bmatrix}$ and $N = \begin{bmatrix} 1 & -5 & 9 \\ 0 & -1 & 1 \\ 0 & 1 & -2 \end{bmatrix}$.

Then $M^{-1} = \begin{bmatrix} 1 & 0 & 0 \\ 2 & 1 & 0 \\ 1 & -1 & 1 \end{bmatrix}$, $N^{-1} = \begin{bmatrix} 1 & -1 & 4 \\ 0 & -2 & -1 \\ 0 & -1 & -1 \end{bmatrix}$, which leads to the representation

$A = \begin{bmatrix} 1 & 0 \\ 2 & 1 \\ 1 & -1 \end{bmatrix}\begin{bmatrix} 1 & -1 & 4 \\ 0 & -2 & -1 \end{bmatrix}$. The mate of $\begin{bmatrix} 1 & 0 \\ 2 & 1 \\ 1 & -1 \end{bmatrix}$ is

$$\begin{bmatrix} 6 & 1 \\ 1 & 2 \end{bmatrix}^{-1}\begin{bmatrix} 1 & 2 & 1 \\ 0 & 1 & -1 \end{bmatrix} = \frac{1}{11}\begin{bmatrix} 2 & -1 \\ -1 & 6 \end{bmatrix}\begin{bmatrix} 1 & 2 & 1 \\ 0 & 1 & -1 \end{bmatrix} = \frac{1}{11}\begin{bmatrix} 2 & 3 & 3 \\ -1 & 4 & -7 \end{bmatrix}$$

and the mate of $\begin{bmatrix} 1 & -1 & 4 \\ 0 & -2 & -1 \end{bmatrix}$ is

$$\begin{bmatrix} 1 & 0 \\ -1 & -2 \\ 4 & -1 \end{bmatrix}\begin{bmatrix} 18 & -2 \\ -2 & 5 \end{bmatrix}^{-1} = \frac{1}{86}\begin{bmatrix} 1 & 0 \\ -1 & -2 \\ 4 & -1 \end{bmatrix}\begin{bmatrix} 5 & 2 \\ 2 & 18 \end{bmatrix} = \frac{1}{86}\begin{bmatrix} 5 & 2 \\ -9 & -38 \\ 18 & -10 \end{bmatrix},$$

so the mate of the given matrix is

$$\frac{1}{11\cdot 86}\begin{bmatrix} 5 & 2 \\ -9 & -38 \\ 18 & -10 \end{bmatrix}\begin{bmatrix} 2 & 3 & 3 \\ -1 & 4 & -7 \end{bmatrix} = \frac{1}{946}\begin{bmatrix} 8 & 23 & 1 \\ 20 & -179 & 239 \\ 46 & 14 & 124 \end{bmatrix}.$$

10 Since R^T is the mate of $S^T = S$, $R^T = R$. RS is symmetric by the definition of a mate. Thus, $RS = (RS)^T = S^T R^T = SR$, as was to be shown.

11 Since $B = BAB = B(AB)^T = BB^TA^T$ and $B = (BA)^TB = A^TB^TB$, the mate of A is divisible on both the left and the right by A^T. Suppose C is a matrix divisible on both the left and the right by A^T and $ACA = A$. The difference $D = B - C$ is divisible on left and right by A^T and satisfies $ADA = 0$. Let $D = QA^T$ and $D = A^TR$. Because $(AQA^T)(AQA^T)^T = AQA^TAQ^TA^T = ADAQ^TA^T = 0Q^TA^T = 0$, $AQA^T = 0$. Therefore, $D^TD = (A^TR)^TD = R^TAD = R^TAQA^T = R^T0 = 0$, so $D = 0$, as was to be shown.

Chapter 8

1

$$\begin{bmatrix} x^2+x+1 & x^2-x+1 \\ 1 & 0 \\ 0 & 1 \end{bmatrix} \sim \begin{bmatrix} x^2+x+1 & -2x \\ 1 & -1 \\ 0 & 1 \end{bmatrix} \sim \begin{bmatrix} x+1 & -2x \\ -\frac{1}{2}x+1 & -1 \\ \frac{1}{2}x & 1 \end{bmatrix} \sim$$

$$\sim \begin{bmatrix} x+1 & 2 \\ -\frac{1}{2}x+1 & -x+1 \\ \frac{1}{2}x & x+1 \end{bmatrix} \sim \begin{bmatrix} 1 & 2 \\ \frac{1}{2}x^2-x+1 & -x+1 \\ -\frac{1}{2}x^2 & x+1 \end{bmatrix} \sim \begin{bmatrix} 1 & 0 \\ \frac{1}{2}x^2-x+1 & -x^2+x-1 \\ -\frac{1}{2}x^2 & x^2+x+1 \end{bmatrix}$$

gives the solution $N = \begin{bmatrix} \frac{1}{2}x^2-x+1 & -x^2+x-1 \\ -\frac{1}{2}x^2 & x^2+x+1 \end{bmatrix}$.

2

$$\begin{bmatrix} 8x^3+1 & 4x^2-1 \\ 1 & 0 \\ 0 & 1 \end{bmatrix} \sim \begin{bmatrix} 2x+1 & 4x^2-1 \\ 1 & 0 \\ -2x & 1 \end{bmatrix} \sim \begin{bmatrix} 2x+1 & -2x-1 \\ 1 & -2x \\ -2x & 4x^2+1 \end{bmatrix} \sim$$

$$\sim \begin{bmatrix} x+\frac{1}{2} & -2x-1 \\ -x+1 & -2x \\ 2x^2-2x+\frac{1}{2} & 4x^2+1 \end{bmatrix} \sim \begin{bmatrix} x+\frac{1}{2} & 0 \\ -x+1 & -4x+2 \\ 2x^2-2x+\frac{1}{2} & 8x^2-4x+2 \end{bmatrix}$$

gives $\begin{bmatrix} 8x^3+1 & 4x^2-1 \end{bmatrix}\begin{bmatrix} -x+1 & -4x+2 \\ 2x^2-2x+\frac{1}{2} & 8x^2-4x+2 \end{bmatrix} = \begin{bmatrix} x+\frac{1}{2} & 0 \end{bmatrix}$, so the gcd is $x+\frac{1}{2} = (8x^3+1)(-x+1)+(4x^2-1)(2x^2-2x+\frac{1}{2})$.

3 $\begin{bmatrix} x+1 & 0 \\ 0 & 2x^2-2 \end{bmatrix}$.

4 The computation in the examples section follows the algorithm for the first five steps. From that point on, the algorithm gives

$$\begin{bmatrix} 1 & 0 & 0 \\ -x^2+7x-13 & x^2-6x+10 & x-4 \\ 2x^2-13x+19 & -2x^2+11x-14 & -x+2 \end{bmatrix} \sim$$

$$\sim \begin{bmatrix} 1 & 0 & 0 \\ x^2-6x+6 & -x^2+5x-4 & -2 \\ 2x^2-13x+19 & -2x^2+11x-14 & -x+2 \end{bmatrix} \sim$$

$$\sim \begin{bmatrix} 1 & 0 & 0 \\ x^2-6x+6 & -x^2+5x-4 & -2 \\ -x+7 & x-6 & -x+6 \end{bmatrix} \sim \begin{bmatrix} 1 & 0 & 0 \\ x+6 & -x-4 & -x^2+6x-2 \\ -x+7 & x-6 & -x+6 \end{bmatrix} \sim$$

$$\sim \begin{bmatrix} 1 & 0 & 0 \\ x+6 & -x-4 & -x^2+6x-2 \\ 13 & -10 & -x^2+5x+4 \end{bmatrix} \sim \begin{bmatrix} 1 & 0 & 0 \\ 6 & -\frac{3}{13}x-4 & \frac{1}{13}x^3-\frac{18}{13}x^2+\frac{74}{13}x-2 \\ 13 & -10 & -x^2+5x+4 \end{bmatrix} \sim$$

$$\sim \begin{bmatrix} 1 & 0 & 0 \\ 1 & -\frac{3}{13}x-\frac{2}{13} & \frac{1}{13}x^3-x^2+\frac{49}{13}x-\frac{46}{13} \\ 13 & -10 & -x^2+5x+4 \end{bmatrix} \sim$$

$$\sim \begin{bmatrix} 1 & 0 & 0 \\ 1 & -\frac{3}{13}x-\frac{2}{13} & \frac{1}{13}x^3-x^2+\frac{49}{13}x-\frac{46}{13} \\ 0 & 3x-8 & -x^3+12x^2-44x+50 \end{bmatrix} \sim$$

$$\sim \begin{bmatrix} 1 & 0 & 0 \\ 0 & -\frac{3}{13}x-\frac{2}{13} & \frac{1}{13}x^3-x^2+\frac{49}{13}x-\frac{46}{13} \\ 0 & 3x-8 & -x^3+12x^2-44x+50 \end{bmatrix} \sim$$

$$\sim \begin{bmatrix} 1 & 0 & 0 \\ 0 & -\frac{3}{13}x-\frac{2}{13} & -\frac{41}{39}x^2+\frac{49}{13}x-\frac{46}{13} \\ 0 & 3x-8 & \frac{28}{3}x^2-44x+50 \end{bmatrix} \sim \begin{bmatrix} 1 & 0 & 0 \\ 0 & -\frac{3}{13}x-\frac{2}{13} & \frac{523}{117}x-\frac{46}{13} \\ 0 & 3x-8 & -\frac{13}{3}x^2-\frac{68}{9}x+50 \end{bmatrix} \sim$$

$$\sim \begin{bmatrix} 1 & 0 & 0 \\ 0 & x-\frac{7670}{6799} & \frac{523}{117}x-\frac{46}{13} \\ 0 & -\frac{624}{523}x^2+\frac{481}{523}x+\frac{3016}{523} & -\frac{13}{3}x^2-\frac{68}{9}x+50 \end{bmatrix} \sim$$

$$\sim \begin{bmatrix} 1 & 0 & 0 \\ 0 & x-\frac{7670}{6799} & \frac{176}{117} \\ 0 & -\frac{624}{523}x^2+\frac{481}{523}x+\frac{3016}{523} & x^2-\frac{35}{3}x+\frac{218}{9} \end{bmatrix} \sim$$

$$\sim \begin{bmatrix} 1 & 0 & 0 \\ 0 & -\frac{7670}{6799} & \frac{176}{117} \\ 0 & -\frac{117}{176}x^3+\frac{604071}{92048}x^2-\frac{1397526}{92048}x+\frac{3016}{523} & x^2-\frac{35}{3}x+\frac{218}{9} \end{bmatrix} \sim$$

$$\sim \begin{bmatrix} 1 & 0 & 0 \\ 0 & 1 & \frac{176}{117} \\ 0 & -\frac{117}{176}x^3+\frac{351}{44}x^2-\frac{507}{16}x+\frac{3523}{88} & x^2-\frac{35}{3}x+\frac{218}{9} \end{bmatrix} \sim$$

$$\sim \begin{bmatrix} 1 & 0 & 0 \\ 0 & 1 & 0 \\ 0 & -\frac{117}{176}x^3+\frac{351}{44}x^2-\frac{507}{16}x+\frac{3523}{88} & x^3-11x^2+36x-36 \end{bmatrix}$$

from which the expected result follows.

5 A simple example is $\begin{bmatrix} \frac{1}{2} & 0 \\ 0 & 2 \end{bmatrix}$. Since this matrix is strongly diagonal but not equal to the strongly diagonal matrix I_2, it is not equivalent to I_2, that is, it is not unimodular. On the other hand, the strongly diagonal matrix of polynomials to which it is equivalent in the wider sense must have the same determinant 1, and the only strongly diagonal 2×2 matrix of polynomials with determinant 1 is I_2, so the matrix is unimodular in the wider sense.

6 As in Chapter 5, if $B = TA$ or $B = AT$ where T is a polynomial tilt, then every $k \times k$ minor of B is either a $k \times k$ minor of A or it is a $k \times k$ minor of A plus a multiple of another $k \times k$ minor of A. Therefore, any common divisor of the $k \times k$ minors of A is a common divisor of the $k \times k$ minors of B and is consequently a common divisor of the $k \times k$ minors of any matrix equivalent to A. A $k \times k$ minor of a strongly diagonal matrix is either zero or the product of k distinct nonzero diagonal entries, so the product of the first k diagonal entries is a common divisor of the $k \times k$ minors.

7 Direct use of the formula gives $(x-3)(x-5)(x-3) + (-1)(-1)(-1) + (-1)(-1)(-1) - (x-3)(-1)(-1) - (x-5)(-1)(-1) - (x-3)(-1)(-1) = x^3 - 11x^2 + 39x - 45 - 2 - x + 3 - x + 5 - x + 3 = x^3 - 11x^2 + 36x - 36$. Alternatively, elimination can be used to reduce to a 2×2 determinant. See also Exercise 5 of Chapter 9.

8 Adding a multiple of row i to row $j \neq i$ in an $m \times n$ matrix A is the same as forming the matrix product AM where M is the $n \times n$ matrix obtained by adding the same multiple of row i to row j of I_n. What is to be shown is that M is unimodular. The ith 1 on the diagonal of M is the only nonzero entry in its row, from which it is clear that $|M| = 1$. Therefore, the strongly diagonal matrix to which M is equivalent must have determinant 1. Since I_n is the only such strongly diagonal matrix of polynomials, $M \sim I_n$. Similarly, adding a multiple of a column to another column is multiplication on the left by a unimodular matrix.

Chapter 9

1 $C = \begin{bmatrix} 1 & 0 \\ 0 & -1 \end{bmatrix}$ (or $C = \begin{bmatrix} -1 & 0 \\ 0 & 1 \end{bmatrix}$). There are many matrices P with $A = P^{-1}CP$. For example, $P = \begin{bmatrix} 1 & 1 \\ 1 & -1 \end{bmatrix}$ has this property.

2 For any square matrix of rational numbers A, $A = I^{-1}AI$ shows that A is similar to itself. If A is similar to B, then $A = P^{-1}BP$ for some invertible P; then $PAP^{-1} = B$, which shows that B is similar to A. Finally, if $A = P^{-1}BP$ and $B = Q^{-1}CQ$, then $A = P^{-1}Q^{-1}CQP = (QP)^{-1}C(QP)$, which shows that A is similar to C.

3 The sum of the three 1×1 principal minors is -7. The sum of the three 2×2 principal minors is $2 + 8 + 2 = 12$, and the single 3×3 minor is 0. Therefore, the characteristic polynomial is $x^3 + 7x^2 + 12x = x(x+3)(x+4)$. Since it has no multiple factors, it is equal to the minimum polynomial of this matrix, so the matrix is diagonalizable. Specifically, the rational canonical form $\begin{bmatrix} 0 & 0 & 0 \\ 0 & -3 & 0 \\ 0 & 0 & -4 \end{bmatrix}$ is a diagonal matrix similar to the given one.

4 The characteristic polynomial is $x^3 + x^2 + x + 1$. For the given A,

$$A^2 = \begin{bmatrix} -1 & 0 & 0 \\ -63 & -10 & 33 \\ -21 & -3 & 10 \end{bmatrix} \quad \text{and} \quad A^3 = \begin{bmatrix} 28 & 5 & -15 \\ 50 & 8 & -27 \\ 69 & 12 & -37 \end{bmatrix}$$

from which $A^3 + A^2 + A + I = 0$ follows. Because of the factorization $x^3 + x^2 + x + 1 = (x+1)(x^2+1)$, the rational canonical form is $\begin{bmatrix} -1 & 0 & 0 \\ 0 & 0 & -1 \\ 0 & 1 & 0 \end{bmatrix}$.

5 The signs are more easily managed if one proves the equivalent formula $|xI + A| = x^n + c_1x^{n-1} + c_2x^{n-2} + \cdots + c_jx^{n-j} + \cdots + c_n$. The determinant $|xI + A|$ is, in the first instance, a sum of $n!$ terms, each of which is 1 or -1 times a product of n entries of $xI + A$. The products containing one or more diagonal entries $x + a_{ii}$ can be further expanded to write $|xI + A|$ as a sum of even more terms, each of which is of the form $\pm Px^j$, where P is a product of $n - j$ entries of A and $0 \leq j \leq n$. Each such term corresponds to a choice of j diagonal positions (the positions contributing x's) followed by the choice of $n - j$ other positions in the $n \times n$ array (contributing a's) in such a way that the set of all n chosen positions includes one from each row and one from each column. For each choice of j diagonal positions, there are exactly $(n - j)!$ ways to choose the other positions, and each such choice gives one of the $(n - j)!$ terms $\pm P$ of the principal $(n - j) \times (n - j)$ minor of A that omits the j chosen diagonal entries. The sign of such a term $\pm P$ is determined by the parity of the number of row or column interchanges that can be used to put all of the n chosen locations on the diagonal, which is the same way that the sign of the term $\pm P$ in its $(n - j) \times (n - j)$ principal minor of A is determined. Therefore, the coefficient of x^j in $|xI + A|$ is the sum c_{n-j} of all $\binom{n}{j} = \binom{n}{n-j}$ of the principal $(n - j) \times (n - j)$ minors of A, as was to be shown.

6 For example, the matrices $\begin{bmatrix} 0 & 0 & 0 & 0 \\ 0 & 0 & 0 & 0 \\ 0 & 0 & 0 & 0 \\ 0 & 0 & 1 & 0 \end{bmatrix}$ and $\begin{bmatrix} 0 & 0 & 0 & 0 \\ 1 & 0 & 0 & 0 \\ 0 & 0 & 0 & 0 \\ 0 & 0 & 1 & 0 \end{bmatrix}$ have elementary divisors x, x, x^2 and x^2, x^2, respectively. The minimum polynomial is x^2 in both cases, but the elementary divisors are different, so the matrices are not similar.

7 As was shown in Section 4, the strongly diagonal matrix equivalent to $xI - A$ has $x^5 - x^3 - x^2 + 1$ as its last diagonal entry and 1 as its other diagonal entries. Clearly 1 and -1 are roots of this polynomial. Division by the factors $x - 1$ and $x + 1$ gives the quotient $x^3 - 1 = (x - 1)(x^2 + x + 1)$. Since $x^2 + x + 1$ is irreducible, it follows that the elementary divisors of A are $x + 1$, $(x - 1)^2$, $x^2 + x + 1$. Therefore, the rational canonical form of A is

$$\begin{bmatrix} -1 & 0 & 0 & 0 & 0 \\ 0 & 1 & 0 & 0 & 0 \\ 0 & 1 & 1 & 0 & 0 \\ 0 & 0 & 0 & 0 & -1 \\ 0 & 0 & 0 & 1 & -1 \end{bmatrix}.$$

8 The definition of the minimum polynomial of a matrix given in Section 6 applies to matrices of numbers, not to matrices of polynomials. (A natural definition can be given, but the minimum polynomial of a matrix of polyonomials in one variable is a polynomial in two variables.) If A is any 3×3 matrix of numbers whose elementary divisors are $x - 2$ and $(x - 2)^2$—for example, if A is $2I$ with the middle entry of the last row changed to 1—then $(x - 2)^2$ is the minumum polynomial of A.

9 Let $A = \begin{bmatrix} 0 & -1 \\ 1 & 0 \end{bmatrix}$ and $B = \begin{bmatrix} i & 0 \\ 0 & -i \end{bmatrix}$. The reduction of $xI - A$ and $xI - B$ to strongly diagonal form using the algorithm gives

$$\begin{bmatrix} 1 & 0 \\ x^2 - x + 1 & 1 \end{bmatrix}(xI - A)\begin{bmatrix} 1 & 0 \\ -x + 1 & 1 \end{bmatrix}\begin{bmatrix} 1 & -1 \\ 0 & 1 \end{bmatrix} = \begin{bmatrix} 1 & 0 \\ 0 & x^2 + 1 \end{bmatrix} =$$

$$= \begin{bmatrix} 1 & 0 \\ -\frac{i}{2}x^2 + \frac{i}{2}x - \frac{1+i}{2} & 1 \end{bmatrix}\begin{bmatrix} 1 & 1 \\ 0 & 1 \end{bmatrix}(xI - B)\begin{bmatrix} 1 & -1 \\ 0 & 1 \end{bmatrix}\begin{bmatrix} 1 & 0 \\ -\frac{x}{2i} & 1 \end{bmatrix}\begin{bmatrix} 1 & 0 \\ \frac{1-i}{2} & 1 \end{bmatrix}\begin{bmatrix} 1 & -2i \\ 0 & 1 \end{bmatrix}.$$

Therefore, $xI - B = M(xI - A)N$ where

$$M = \begin{bmatrix} 1 & -1 \\ 0 & 1 \end{bmatrix}\begin{bmatrix} 1 & 0 \\ \frac{i}{2}x^2 - \frac{i}{2}x + \frac{1+i}{2} & 1 \end{bmatrix}\begin{bmatrix} 1 & 0 \\ x^2 - x + 1 & 1 \end{bmatrix}$$

$$= \begin{bmatrix} -(1 + \frac{i}{2})x^2 + (1 + \frac{i}{2})x - \frac{1}{2} - \frac{i}{2} & -1 \\ (1 + \frac{i}{2})x^2 - (1 + \frac{i}{2})x + \frac{3}{2} + \frac{i}{2} & 1 \end{bmatrix},$$

so $R_1 = B^2M_0 + BM_1 + M_2 = \begin{bmatrix} -i & -1 \\ -i & 1 \end{bmatrix}$.

10 The matrix given by the method of the preceding exercise is $\begin{bmatrix} \frac{1}{2}-\frac{\sqrt{5}}{2} & -1 \\ -\frac{1}{2}-\frac{\sqrt{5}}{2} & 1 \end{bmatrix}$. (One need not do the whole calculation. The form of the computation is near enough to the one of the preceding exercise that it can be seen to yield a matrix of the form $R_1 = \begin{bmatrix} r & -1 \\ s & 1 \end{bmatrix}$. The values of r and s are then easy to find using $R_1 \begin{bmatrix} 1 & 1 \\ 1 & 2 \end{bmatrix} = \begin{bmatrix} (3+\sqrt{5})/2 & 0 \\ 0 & (3-\sqrt{5})/2 \end{bmatrix} R_1$.)

11 If there were a counterexample—that is, a polynomial $f(x)$ of degree n and $n+1$ distinct numbers $a_1, a_2, \ldots, a_{n+1}$ such that $f(a_i) = 0$ for all i—then a smaller counterexample could be constructed as follows. Because a polynomial of degree 0 has no roots, $n > 0$. Let $f(x)$ be divided by $x - a_1$ to find $f(x) = q(x)(x - a_1) + r$, where r is a number. Substitution of a_1 for x in this equation gives $0 = q(a_1)(a_1 - a_1) + r$, that is, $r = 0$. Thus $f(x) = q(x)(x - a_1)$ where $q(x)$ is a polynomial of degree $n - 1$. This is a smaller counterexample because $q(x)$ has the n distinct roots $a_2, a_3, \ldots, a_{n+1}$. ($0 = f(a_i) = q(a_i)(a_i - a_1)$ can be divided by $a_i - a_1$ when $i > 1$ to find $0 = q(a_i)$.) Therefore, by the principle of infinite descent, there can be no counterexample.

12 Say $f(x) = g(x)q(x)$ where $q(x)$ is irreducible. If $q(x)$ is a multiple of $g(x)$, write $f(x) = g(x)^2 q_1(x)$; if $q_1(x)$ is a multiple of $g(x)$, write $f(x) = g(x)^3 q_2(x)$, and so forth. In this way, $f(x)$ can be written in the form $f(x) = g(x)^j r(x)$ for some integer $j > 0$, where $r(x)$ is not a multiple of $g(x)$. Let $d(x)$ be the gcd of $g(x)$ and $r(x)$. Since $g(x)$ is irreducible, $d(x)$ can only be 1 or $g(x)$. Since $r(x)$ is not a multiple of $g(x)$, $d(x) = 1$. Therefore, $1 = g(x)s(x) + r(x)t(x)$ for some polynomials $s(x)$ and $t(x)$. Since $\big(g(x)s(x)\big)^j = \big(1 - r(x)t(x)\big)^j = 1 +$ terms that are multiples of $r(x)$, any common divisor of $g(x)^j$ and $r(x)$ must be a divisor of 1. In other words, $f(x) = g(x)^j r(x)$ gives $f(x)$ as a product of relatively prime factors. Thus, $f(x)$ is either a power of $g(x)$ (if $r(x) = 1$) or it is a product of two relatively prime factors of positive degree (otherwise).

13 Each diagonal entry of C_1 is either 1 or a power of an irreducible monic polynomial. Let $p_1(x), p_2(x), \ldots, p_k(x)$ be a list of the *distinct* irreducible monic polynomials which divide diagonal entries of D, and let e_{ij} be the exponent of $p_i(x)$ in the jth diagonal entry of C_1. There are nk numbers e_{ij}, where n is the number of rows and columns in C_1, but at most n of the e_{ij} are nonzero. For each i, let the exponents e_{ij} be rearranged in nondecreasing order, and let $f_{i1} \le f_{i2} \le \cdots \le f_{in}$ be the e's reordered in this way. By Proposition 1 of Section 5, and by the fact that a diagonal matrix is equivalent to any matrix obtained from it by reordering its diagonal entries, C_1 is equivalent to the matrix whose jth diagonal entry is $p_1(x)^{f_{1j}} p_2(x)^{f_{2j}} \cdots p_k(x)^{f_{kj}}$. But this matrix, call it D, is strongly diagonal because the sequences of the fs are nondecreasing. Therefore, D is the unique strongly diagonal matrix equivalent to C_1. The kn exponents f in the expression of the diagonal entries of D are the same—except for their order—as the kn exponents e, so D determines the diagonal entries of C_1, except for their order. If $C_1 \sim C_2$, then D is the strongly diagonal matrix equivalent to C_2, and D determines the diagonal entries of C_2 in the same way, from which the desired conclusion follows.

14 Let r be a rational root of $f(x) = x^n + c_1 x^{n-1} + \cdots + c_n$, where $c_1, c_2, \ldots, c_n$ are integers. There exist relatively prime integers p and q such that $r = p/q$. The equation $p^n = -c_1 p^{n-1} q - c_2 p^{n-2} q^2 - \cdots - c_n q^n$ shows that p^n is a multiple of q. Because p and q are relatively prime, there are integers a and b such that $ap = bq + 1$. Then $a^n p^n = (1 + bq)^n$ is a multiple of q, which implies that 1 is a multiple of q. Therefore, $q = \pm 1$, so $r = \pm p$ is an integer, as was to be shown.

15 Suppose $f(x)g(x)$ is a multiple of $x - a$. The only monic polynomials of which $x - a$ is a multiple are itself and 1, because the multiplier can only have degree 0 or 1. Therefore, the gcd of $x - a$ and $f(x)$ is either $x - a$ or 1. If it is $x - a$, then $f(x)$ is a multiple of $x - a$. If it is 1, then $1 = a(x)f(x) + b(x)(x - a)$ for some polynomials $a(x)$ and $b(x)$, so $g(x) = a(x)f(x)g(x) + b(x)(x - a)g(x)$ is a multiple of $x - a$ because it is a sum of two terms, each of which is a multiple of $x - a$. Thus, either $f(x)$ or $g(x)$ is a multiple of $x - a$ as was to be shown.

Suppose $\prod(x - a_i)$ is a multiple of $f(x)$. It will be shown that $f(x)$ is equal to a nonzero rational number times the product of a subset of the factors $x - a_i$. Suppose $f(x)g(x) = \prod(x - a_i)$ and suppose $f(x)$ does not have the stated form. The number of factors on the right must be positive, because an equation of the form $f(x)g(x) = 1$ implies that both factors on the left are nonzero rational numbers, so $f(x)$ has the stated form, contrary to assumption. Then, since $x - a_1$ divides either $f(x)$ or $g(x)$, division by $x - a_1$ would give another equation of the same form $f_1(x)g_1(x) = \prod_{i>1}(x - a_i)$ in which the number of factors on the right side is reduced by one and $f_1(x)$ does not have the stated form (because either $f_1(x) = f(x)$ or $f_1(x)(x - a_1) = f(x)$, so $f(x)$ would have the stated form if $f_1(x)$ did). This shows by the principle of infinite descent—if the existence of a counterexample implies the existence of a smaller counterexample, there can be no counterexample—that $f(x)$ must have the stated form.

16 By its definition, $A^{(j)}$ is 1 in positions where the row index is one greater than the column index and 0 in all other positions. Its square is 1 in positions where the row index is two greater than the column index and 0 elsewhere. In general, the kth power of $A^{(j)}$ is 1 in positions (if any) where the row index is k greater than the column index and 0 elsewhere. It follows that $g(A^{(j)}) = 0$ if and only if the coefficient of x^i in $g(x)$ is zero for $i = 0, 1, \ldots, j - 1$, so the minimum polynomial of $A^{(j)}$ is x^j. Therefore, the strongly diagonal matrix equivalent to $xI - A^{(j)}$ has as its last diagonal entry x^j. Since its determinant is monic of degree j, its other diagonal entries must be 1, as was to be shown.

17 Let B be the $j \times j$ matrix which is 1 in positions where the row index is one greater than the column index and 0 elsewhere. The result proved in the preceding exercise implies that $xI - B$ is equivalent to the matrix obtained by replacing the last diagonal entry of I_j with x^j. Therefore $(x - a)I - B$, where a is a rational number, is equivalent to the matrix obtained by replacing the last diagonal entry of I_j with $(x - a)^j$. If A is the 1×1 matrix whose only entry is a, then $A^{(j)} = aI + B$. Thus, $xI - A^{(j)} = (x - a)I - B$ is equivalent to the required matrix.

18 The square of $A^{(j)}$ is made up of $n \times n$ blocks. The diagonal blocks are A^2, the blocks just below the diagonal (where $A^{(j)}$ has I) are $2A$, the blocks just below them—that is, the blocks where, for some $i > 1$, the row index is between $ni + 1$ and $ni + n$ and the column index is between $n(i - 2) + 1$ and $n(i - 2) + n$—are I, while all other blocks are zero. The cube of $A^{(j)}$ has A^3 in the diagonal blocks, $3A^2$ in the blocks below the diagonal, $3A$ in the blocks below them, I in the blocks below them, and zeros in all other blocks. In general, the matrix obtained by raising $A^{(j)}$ to the power p consists of $n \times n$ blocks; blocks above the diagonal are zero, blocks on the diagonal are A^p, and blocks k steps below the diagonal are $\binom{p}{k}A^{p-k}$, where $\binom{p}{k}$ denotes the binomial coefficient. This formula is easy to prove by mathematical induction.

Otherwise stated, the blocks k steps below the diagonal contain the coefficient of h^k in $g(A + hI)$ when $g(x) = x^p$, while blocks above the diagonal are zero. The cases $g(x) = x^p$ of this formula imply the case of a general polynomial in x with rational coefficients. Thus, given a polynomial $g(x)$, the blocks of $g(A^{(j)})$ that are k steps below the diagonal ($k = 0, 1, \ldots, j - 1$) contain $g_k(A)$, where the polynomials $g_k(x)$ are the coefficients of $g(x + h)$ as a polynomial in h, that is, $g(x + h) = g_0(x) + g_1(x)h + g_2(x)h^2 + \cdots$. (Of course $g_0(x) = g(x)$ and $g_1(x)$ is the derivative of $g(x)$.)

Let $g(x)$ be the minimum polynomial of $A^{(j)}$, and let it be written in the form $g(x) = f(x)^l q(x)$ where l is an integer and $q(x)$ is a polynomial that is not divisible by $f(x)$ (see the answer to Exercise 12). Then $g(A + hI) = f(A + hI)^l q(A + hI) = \left(f'(A)h + \cdots\right)^l \left(q(A) + \cdots\right) = f'(A)^l q(A)h^l +$ terms containing h^{l+1}, where $f'(A)$ is the coefficient of h in $f(A + hI)$. Because $f'(x)$ has lower degree than $f(x)$, the gcd of $f(x)$ and $f'(x)$ has lower degree than $f(x)$, which implies, because $f(x)$ is irreducible, that the gcd is 1. Therefore, there exist polynomials $n_1(x)$ and $n_2(x)$ such that $n_1(x)f(x) + n_2(x)f'(x) = 1$. Therefore, the matrix $n_2(A)$ satisfies $n_2(A)f'(A) = I$; in particular, $f'(A)$ is invertible. Since $q(A) \neq 0$ (by assumption, $q(x)$ is not divisible by $f(x)$), it follows that the coefficient $f'(A)^l q(A)$ of h^l in $g(A + hI)$ is not zero. If l had any of the values 0, $1, \ldots, j - 1$, it would follow that the blocks l steps below the diagonal of $g(A^{(j)})$ were nonzero, contrary to assumption. Therefore, $l \geq j$, which shows that $f(x)^j$ is a factor of $g(x)$. Since $g(x)$ is a factor of the characteristic polynomial of $A^{(j)}$ (the Cayley-Hamilton theorem), and since this characteristic polynomial has the same degree as $f(x)^j$, it follows that $g(x) = f(x)^j$, as was to be shown.

Chapter 10

1

(a) $3\begin{bmatrix} \frac{4}{5} & \frac{2}{5} \\ \frac{2}{5} & \frac{1}{5} \end{bmatrix} - 2\begin{bmatrix} \frac{1}{5} & -\frac{2}{5} \\ -\frac{2}{5} & \frac{4}{5} \end{bmatrix}$ (b) $13\begin{bmatrix} \frac{9}{13} & \frac{6}{13} \\ \frac{6}{13} & \frac{4}{13} \end{bmatrix} + 26\begin{bmatrix} \frac{4}{13} & -\frac{6}{13} \\ -\frac{6}{13} & \frac{9}{13} \end{bmatrix}$

(c) $-7\begin{bmatrix} \frac{1}{14} & \frac{2}{14} & \frac{3}{14} \\ \frac{2}{14} & \frac{4}{14} & \frac{6}{14} \\ \frac{3}{14} & \frac{6}{14} & \frac{9}{14} \end{bmatrix} + 7\begin{bmatrix} \frac{13}{14} & -\frac{2}{14} & -\frac{3}{14} \\ -\frac{2}{14} & \frac{10}{14} & -\frac{6}{14} \\ -\frac{3}{14} & -\frac{6}{14} & \frac{5}{14} \end{bmatrix}$

(d) $\mathrm{tr}(S) = -3$ and $\det(S) = 3^3 \cdot 19^2$ leads to the conclusion that if the roots are rational then the characteristic polynomial is $(x+3)(x+57)(x-57)$. Because $3 \cdot 57 - 3 \cdot 57 - 57 \cdot 57$ is the sum of the 2×2 principal minors of S, this polynomial is in fact the characteristic polynomial of S. The first column of S^2 is $\begin{bmatrix} 2169 \\ -1080 \\ -1080 \end{bmatrix}$. Therefore, the first column of $(S - 57I)(S + 57I) = S^2 - 57^2 I$ has -1080 in all three entries. Therefore, the P associated with the root -3 of the characteristic polynomial of S is AA^M, where A is the column matrix whose entries are all 1. The first column of $(S+3I)(S+57I) = S^2 + 60S + 171I$ has entries 3840, -480, -3360. Thus, the P associated to the root 57 is AA^M where A is the column matrix with entries 8, -1, -7. These observations lead to the easily verified conclusion

$$S = -3\begin{bmatrix} \frac{1}{3} & \frac{1}{3} & \frac{1}{3} \\ \frac{1}{3} & \frac{1}{3} & \frac{1}{3} \\ \frac{1}{3} & \frac{1}{3} & \frac{1}{3} \end{bmatrix} - 57\begin{bmatrix} \frac{12}{114} & -\frac{30}{114} & \frac{18}{114} \\ -\frac{30}{114} & \frac{75}{114} & -\frac{45}{114} \\ \frac{18}{114} & -\frac{45}{114} & \frac{27}{114} \end{bmatrix} + 57\begin{bmatrix} \frac{64}{114} & -\frac{8}{114} & -\frac{56}{114} \\ -\frac{8}{114} & \frac{1}{114} & \frac{7}{114} \\ -\frac{56}{114} & \frac{7}{114} & \frac{49}{114} \end{bmatrix}.$$

2

$$P = \begin{bmatrix} \frac{a^2}{a^2+b^2} & \frac{ab}{a^2+b^2} \\ \frac{ab}{a^2+b^2} & \frac{b^2}{a^2+b^2} \end{bmatrix} \quad \text{and} \quad Q = \begin{bmatrix} \frac{b^2}{a^2+b^2} & -\frac{ab}{a^2+b^2} \\ -\frac{ab}{a^2+b^2} & \frac{a^2}{a^2+b^2} \end{bmatrix}.$$

3 Let the matrices of the given partition be P and Q. Let A be any nonzero matrix of the form $A = PX$ where X is a 4×1 matrix. Let A^M be the mate of A and let $P_1 = AA^M$, $P_2 = P - P_1$. For example, if X has entries 1, 0, 0, 0, then A is the first column of P and the decomposition of P is

$$P_1 + P_2 = \begin{bmatrix} \frac{3}{5} & \frac{1}{5} & \frac{1}{5} & -\frac{2}{5} \\ \frac{1}{5} & \frac{1}{15} & \frac{1}{15} & -\frac{2}{15} \\ \frac{1}{5} & \frac{1}{15} & \frac{1}{15} & -\frac{2}{15} \\ -\frac{2}{5} & -\frac{2}{15} & -\frac{2}{15} & \frac{4}{15} \end{bmatrix} + \begin{bmatrix} 0 & 0 & 0 & 0 \\ 0 & \frac{1}{3} & \frac{1}{3} & \frac{1}{3} \\ 0 & \frac{1}{3} & \frac{1}{3} & \frac{1}{3} \\ 0 & \frac{1}{3} & \frac{1}{3} & \frac{1}{3} \end{bmatrix}.$$

Similarly,

$$Q = Q_1 + Q_2 = \begin{bmatrix} \frac{2}{5} & -\frac{1}{5} & -\frac{1}{5} & \frac{2}{5} \\ -\frac{1}{5} & \frac{1}{10} & \frac{1}{10} & -\frac{1}{5} \\ -\frac{1}{5} & \frac{1}{10} & \frac{1}{10} & -\frac{1}{5} \\ \frac{2}{5} & -\frac{1}{5} & -\frac{1}{5} & \frac{2}{5} \end{bmatrix} + \begin{bmatrix} 0 & 0 & 0 & 0 \\ 0 & \frac{1}{2} & -\frac{1}{2} & 0 \\ 0 & -\frac{1}{2} & \frac{1}{2} & 0 \\ 0 & 0 & 0 & 0 \end{bmatrix}.$$

That $I_4 = P_1 + P_2 + Q_1 + Q_2$ is an orthogonal partition of unity follows from the fact that each matrix is of the form AA^M for a 4×1 matrix A, namely,

$$A = \begin{bmatrix} 3 \\ 1 \\ 1 \\ -2 \end{bmatrix}, \begin{bmatrix} 0 \\ 1 \\ 1 \\ 1 \end{bmatrix}, \begin{bmatrix} 2 \\ -1 \\ -1 \\ 2 \end{bmatrix}, \begin{bmatrix} 0 \\ 1 \\ -1 \\ 0 \end{bmatrix},$$

respectively and the fact that the transpose of any one of these matrices times any other of these matrices is 0. The fact that $P_1 + P_2 + Q_1 + Q_2$ is an orthogonal partition of unity implies that $P + Q$ is.

6 Let U be the matrix in the Examples section for which $U^T U = I$ and $U^T ST =$ diagonal. Let (x, y) and (z, w) be two pairs of variables connected by the linear substitution whose matrix of coefficients is U. Specifically, let $X = UZ$ where $X = \begin{bmatrix} x \\ y \end{bmatrix}$ and $Z = \begin{bmatrix} z \\ w \end{bmatrix}$. Then the quadratic forms $x^2 + y^2 = X^T X = Z^T U^T U Z = Z^T Z$ and $x^2 + 2xy + 2y^2 = X^T S X = Z^T U^T S U Z$ become $z^2 + w^2$ and $\frac{3+\sqrt{5}}{2} z^2 + \frac{3-\sqrt{5}}{2} w^2$, respectively, in zw-coordinates.

The first equation says that the distance between two points in the xy-plane is the same as the distance between the two corresponding points in the zw-plane. This implies, in particular, that the circle of radius r, $x^2 + y^2 = r^2$, in xy-coordinates coincides with the circle of radius r in zw-coordinates. The curves $x^2 + 2xy + 2y^2 =$ constant in xy-coordinates correspond to the curves $\frac{3+\sqrt{5}}{2} z^2 + \frac{3-\sqrt{5}}{2} w^2 =$ constant in zw-coordinates. But the geometry of this latter set of curves is clear. Because both $\frac{3+\sqrt{5}}{2}$ and $\frac{3-\sqrt{5}}{2}$ are positive, the curves form a family of ellipses whose axes of symmetry are the z- and w-axes. The point where $z = 1$, $w = 0$ is on the circle of radius 1 and on the ellipse $\frac{3+\sqrt{5}}{2} z^2 + \frac{3-\sqrt{5}}{2} w^2 = \frac{3+\sqrt{5}}{2}$. The other points of the circle lie *inside* this ellipse, so the minor axis of the ellipse is the line $w = 0$. In the same way, the major axis is the line $z = 0$. In xy-coordinates, the minor axis $w = 0$ is $bx + dy = 0$ (because $Z = U^T X$) and the major axis is $ax + cy = 0$.

7 The characteristic polynomial of S is $x^3 - 82x^2 + 1935x - 12150$, which has the factorization $(x - 10)(x - 27)(x - 45)$. The first column of $(S - 10I)(S - 27I)$ has as its entries 224, -280, -112. When the common factor 56 is removed, these are 4, -5, -2, so the P associated with the root 45 of the minimum polynomial is

$$P_1 = \begin{bmatrix} 4 \\ -5 \\ -2 \end{bmatrix} \begin{bmatrix} 45 \end{bmatrix}^{-1} \begin{bmatrix} 4 & -5 & -2 \end{bmatrix} = \begin{bmatrix} \frac{16}{45} & -\frac{20}{45} & -\frac{8}{45} \\ -\frac{20}{45} & \frac{25}{45} & \frac{10}{45} \\ -\frac{8}{45} & \frac{10}{45} & \frac{4}{45} \end{bmatrix}.$$

The other two matrices in the partition of unity can be found in the same way, leading to the conclusion that the spectral representation of S is

$$S = 45 \begin{bmatrix} \frac{16}{45} & -\frac{20}{45} & -\frac{8}{45} \\ -\frac{20}{45} & \frac{25}{45} & \frac{10}{45} \\ -\frac{8}{45} & \frac{10}{45} & \frac{4}{45} \end{bmatrix} + 27 \begin{bmatrix} \frac{4}{9} & \frac{4}{9} & -\frac{2}{9} \\ \frac{4}{9} & \frac{4}{9} & -\frac{2}{9} \\ -\frac{2}{9} & -\frac{2}{9} & \frac{1}{9} \end{bmatrix} + 10 \begin{bmatrix} \frac{1}{5} & 0 & \frac{2}{5} \\ 0 & 1 & 0 \\ \frac{2}{5} & 0 & \frac{4}{5} \end{bmatrix}.$$

In accordance with the construction of Section 10, let $C = \begin{bmatrix} 4 & 2 & 1 \\ -5 & 2 & 0 \\ -2 & -1 & 2 \end{bmatrix}$. Then $C^T C = \begin{bmatrix} 45 & 0 & 0 \\ 0 & 9 & 0 \\ 0 & 0 & 5 \end{bmatrix}$ leads to $V = \begin{bmatrix} \sqrt{5}/15 & 0 & 0 \\ 0 & 1/3 & 0 \\ 0 & 0 & \sqrt{5}/5 \end{bmatrix}$ and to $U = CV$ as the matrix of coefficients of the desired linear substitution. Thus, the linear substitution

$$x = \frac{4\sqrt{5}}{15}u + \frac{2}{3}v + \frac{\sqrt{5}}{5}w$$

$$y = -\frac{\sqrt{5}}{3}u + \frac{2}{3}v$$

$$z = -\frac{2\sqrt{5}}{15}u - \frac{1}{3}v + \frac{2\sqrt{5}}{5}w$$

transforms $x^2 + y^2 + z^2$ to $u^2 + v^2 + w^2$ and transforms $30x^2 + 37y^2 + 15z^2 - 16xy - 20xz + 8yz$ to $45u^2 + 27v^2 + 10w^2$. (Note that irrational numbers are needed for this problem, even though S has rational eigenvalues.)

With respect to the coordinates on xyz-space given by u, v, and w, the surfaces in question are ellipsoids, and their axes of symmetry are the coordinate axes, that is, the lines of intersection of the three plane $u = 0$, $v = 0$, $w = 0$.

8 Beginning with $+-+$, rules (1) and (3) imply that only the last sign can change, so the pattern can only change to $+--$. Then rules (2) and (3) imply that only the middle sign can change, and the pattern becomes $++-$. By rules (2) and (3) the next change must be to $+++$, and no further change is possible. Starting with $+-+-$, the next patterns must be $+-++$, $+--+$, but the following pattern can be either $++-+$ or $+---$, after which the remaining patterns are necessarily $++--$, $+++-$, $++++$.

9 Let $m = \deg g$. If $\deg f \geq m$, the solution of the problem for $f(x)$ can be reduced to the solution of the same problem for a polynomial of lower degree in the following way. Let a be the ratio of the leading coefficient of $f(x)$ to the leading coefficient of $g(x)$ and let $j = \deg f - \deg g$. Then $f_1(x) = f(x) - ax^j g(x)$ has lower degree than $f(x)$, and each solution $q(x), r(x)$ of $f(x) = q(x)g(x) + r(x)$ in which $\deg r < m$ corresponds via $q_1(x) = q(x) - ax^j$, $r_1(x) = r(x)$ to one and only one solution $q_1(x), r_1(x)$ of $f_1(x) = q_1(x)g(x) + r_1(x)$ in which $\deg r_1 < m$. When $\deg f < m$, there is a unique solution $q(x) = 0$ and $r(x) = f(x)$, because $q(x) \neq 0$ and $\deg r < m$ imply $\deg(q(x)g(x) + r(x)) \geq m > \deg f$, so $q(x)g(x) + r(x) \neq f(x)$. Note that this proof in fact gives an algorithm for finding the solution.

10 As in the preceding exercise, this statement can be proved by the principle of infinite descent, that is, by proving that the existence of a counterexample would imply the existence of a smaller counterexample. A counterexample would be a polynomial of degree n and a list of $n+2$ numbers $a_1, a_2, \ldots, a_{n+2}$ such that the numbers $f(a_1), f(a_2), \ldots, f(a_{n+2})$ are all nonzero and alternate in sign. Because a polynomial of degree 0 does not change sign, $n > 0$. Let $f'(x)$ be the coefficient of h in $f(x+h)$. Then $f'(x)$ has degree $n-1$, and a smaller counterexample can be found using:

Lemma. *Let $f(x)$ be a polynomial of any degree and let a and b be numbers. If $f(a) < f(b)$ there is a number c in the interval $a \leq c \leq b$ such that $f'(c) > 0$, and if $f(a) > f(b)$ there is a number c in the interval $a \leq c \leq b$ such that $f'(c) < 0$.*

This statement is a basic fact of calculus. It can be proved without calculus in the following way.

Because changing $f(x)$ to $-f(x)$ changes $f'(x)$ to $-f'(x)$, the first of the two statements in the lemma implies the second, so it will suffice to prove the first. By the definition of $f'(x)$, there is a polynomial g in two variables such that $f(x+h) = f(x) + f'(x)h + g(x,h)h^2$. Let B be a number such that $|g(x,h)| \leq B$ whenever x and h lie in the ranges $a \leq x \leq b$ and $0 \leq h \leq b-a$. Let the interval $a \leq x \leq b$ be partitioned into subintervals, all of which have length at most $\frac{f(b)-f(a)}{2B(b-a)}$. Then at least one of the endpoints c of the subintervals must satisfy $f'(c) > 0$ for the following reason. Assume $f'(c) \leq 0$ for all endpoints c, and let c and d be two successive endpoints. Then

$$f(d) = f(c) + f'(c)(d-c) + g(c, d-c)(d-c)^2 \leq f(c) + B(d-c)^2$$
$$\leq f(c) + (d-c)B\frac{f(b)-f(a)}{2B(b-a)} = f(c) + (d-c)\frac{f(b)-f(a)}{2(b-a)}.$$

In other words, on any one subinterval the increase in the value of $f(x)$ is at most $\frac{f(b)-f(a)}{2(b-a)}$ times the length of the subinterval. Thus the total increase in $f(x)$ over all the subintervals partitioning $a \leq x \leq b$ is at most $\frac{f(b)-f(a)}{2(b-a)}$ times the total length $b-a$ of the subintervals. Since this implies $f(b) - f(a) \leq \frac{f(b)-f(a)}{2(b-a)}(b-a) = \frac{f(b)-f(a)}{2}$, it is impossible, as was to be shown.

11 Straightforward computation gives $\det(xI - S) = x^3 - 2x^2 - 10x + 1$. If the characteristic polynomial is not the minimum polynomial of a matrix, the characteristic polynomial is divisible by the square of a polynomial. That this is not the case in the present example follows from the observation that the chararacteristic polynomial changes sign between -3 and -2, between 0 and 1, and between 4 and 5, so it has no multiple roots. The sequence of polynomials generated by the algorithm of Section 6 is $s_3(x) = x^3 - 2x^2 - 10x + 1$, $s_2(x) = 3x^2 - 4x - 10$, $s_1(x) = \frac{68}{9}x + \frac{11}{9}$, $s_0 = \frac{42885}{4624}$.

12 If $A_1, A_2, \ldots, A_m$ are real numbers, then

$$g(x) = A_1\frac{f(x)}{x-a_1} + A_2\frac{f(x)}{x-a_2} + \cdots + A_m\frac{f(x)}{x-a_m}$$

is a polynomial in x of degree less than m in which the coefficient g_1 of x^{m-1} is $A_1 + A_2 + \cdots + A_m$. If $g(x)$ is zero, then, since a_1 is a root of $f(x)/(x-a_i)$ for $i > 1$, it is also a root of $A_1 f(x)/(x-a_1)$; since this polynomial has degree less than m and has the $m-1$ roots $a_2, a_3, \ldots, a_m$, $A_1 f(x)/(x-a_1) = 0$ in this case, that is, $A_1 = 0$. In the same way, $g(x) = 0$ implies $A_i = 0$ for all i. Therefore, the $m \times m$ matrix which expresses the m coefficients of $g(x)$ in terms of $A_1, A_2, \ldots, A_m$ is not a zero divisor, so it is invertible, that is, every polynomial of degree less than m can be expressed in this way for one and only one set of values of $A_1, A_2, \ldots, A_m$. (This expression for g becomes what is called the partial fractions representation of $g(x)/f(x)$ when it is divided by $f(x)$.) Setting $x = a_i$ in the formula for g gives $g(a_i) = A_i f'(a_i)$, because $f(x)/(x-a_j)$ is zero at $x = a_i$ if $i \neq j$ and is, by the very definition, $f'(a_i)$ when $i = j$. Thus, $g_1 = \frac{g(a_1)}{f'(a_1)} + \frac{g(a_2)}{f'(a_2)} + \cdots + \frac{g(a_m)}{f'(a_m)}$. On the other hand, as was shown in Section 6, $g_1 = \operatorname{tr}\left(g(S)h(S)\right)$. Now $h(S)$ is the inverse of $s_{m-1}(S)$, so $h(S) = \frac{1}{s_{m-1}(a_1)}P_1 + \frac{1}{s_{m-1}(a_2)}P_2 + \cdots + \frac{1}{s_{m-1}(a_m)}P_m$, and the trace of $g(S)h(S)$ is $g_1 = \frac{g(a_1)}{s_{m-1}(a_1)}r_1 + \frac{g(a_2)}{s_{m-1}(a_2)}r_2 + \cdots + \frac{g(a_m)}{s_{m-1}(a_m)}r_m$. The agreement of these two formulas for g_1 for all choices of g implies (because g can be chosen to be zero at all but one of the a's) that $\frac{1}{f'(a_i)} = \frac{r_i}{s_{m-1}(a_i)}$ for all i. Thus, $s_{m-1}(x)$ has the stated values at $a_1, a_2, \ldots, a_m$. Any other polynomial of degree less than m with these values differs from $s_{m-1}(x)$ by a polynomial of degree $m-1$ at most that has m zeros, which means that it is identical to $s_{m-1}(x)$.

Index